LYLE'S LAWS

Reflections on Ethics, Engineering, and Everything Else

Lyle D. Feisel

BROOKLYN
RIVER
PRESS

Published in the United States by
Brooklyn River Press
New York

ISBN 978-0-9882675-0-3

COVER: A fractal pattern is a marvel of both simplicity and complexity, whether
viewed from nearby or from a distance. So, too, are Lyle's Laws, each one offering gems
of insight and wisdom in a quick read or in an extended contemplation.

Contents

Introduction

I T WAS 1969 AND WE WERE STANDING on the bridge of a freighter, SS *Hong Kong Bear*, bound from San Francisco to Yokohama. A Japanese fishing boat had appeared almost dead ahead, and since we were steaming at about twenty knots, we were approaching each other at a good clip. The first mate turned to me and said, "Watch this guy. It could be interesting." He was right. As we approached the fishing boat, it suddenly turned and cut directly in front of our ship. We all waited for the collision. Fortunately, it did not come. The boat reappeared to starboard and dropped rapidly astern as we pushed on toward Japan and the fisherman went back to pursuing the wily tuna or whatever it was that he pursued. What was that all about? It's a matter of accumulation and transfer.

In some parts of Asia, there was—and probably still is—a belief that a boat sailing through the water accumulates a trail of spirits, generally of the evil variety, each one glomming onto the last one in line until they create an impossible burden of bad luck. One way to get rid of these unwanted hitchhikers is to force another boat to cut across your wake, whereupon the spirits are transferred from the crossee to the crosser. The closer the encounter, the more of the spirits will be transferred. This has caused some anxious moments for the captains of freighters plying Asian waters, when a fisherman, down on his luck and blaming the spirits hanging onto his fantail, suddenly cuts beneath the freighter's bow, just as that one did with *Hong Kong Bear*. We were lucky. And so was the fisherman. At least he kept his boat; we don't know if the fishing improved or not.

People, as they cruise through life, also accumulate a lot of "stuff." Of course they accumulate some material possessions and, with any luck at all, some measure of financial wealth to see them through their retirement years. More importantly, however, they accumulate a wealth of experience that helps them deal with the various situations that continue to occur in their lives. And maybe some of that experience could be of use to someone else. Is there a way to pass that on? Well, maybe so.

When I was an undergraduate student, I had the good fortune to acquire a grade point average high enough to qualify me for membership in the Tau Beta Pi Association, an engineering honor society. I was moderately active with Tau Beta Pi after graduation and was even honored to receive an award from the association. Then, when I retired, I received a letter from the editor of *The Bent*, Tau Beta Pi's magazine, offering an opportunity to do some writing for the publication. He suggested that I could write a series of "Lyle's Laws" that would give me the opportunity to reflect upon my experiences and perhaps find something of interest and even value to his readers. Is that a deal or what? It sounded good to me so, over the next ten years, I wrote a series of columns—forty in all.

This volume is a collection of those forty laws. When I started, I had no intention of writing a book. As the laws started to accumulate, however, there were many readers who expressed a desire to have all of them in one handy volume, so, now that the series is completed, the collection is being published. The process has given me the incentive to revisit the various laws and reconsider their validity and clarity. There has been some minor editing of the original manuscripts but surprisingly little change in the substance of the laws. This must mean that I did a good job in the first instance or else that I am rigid in my thinking. I hope it is the former.

I claim no originality in most of the laws. The principles expounded are pretty basic and are well known by anyone who has given a little thought to the issues. What I have tried to do is present the laws in a way that will help people see these principles in a slightly different way and perhaps broaden their range of applicability. While the columns were written for an audience made up primarily of engineers, I have attempted to make them applicable to non-engineering activities and hence useful to people who are not engineers. And I think I have succeeded. Over the past few years, we have given reprint permission to business executives, musicians, and economists as well as our engineering colleagues. In addition, many laws offer some pertinent advice on life outside the workplace. And outside the workplace is, indeed, where we spend most of our time.

There is nothing technical in the laws, but they do often refer to—and offer advice on—the engineering process. Non-engineer readers should pick up a measure of insight into that process and perhaps see how engineering thinking can be applied to their own work or personal situations. Certainly our society is heavily influenced by technical considerations and knowing a little more about how engineers think should be useful for anyone.

Referring to my writings as "laws" does trouble me a bit. Webster defines a law as "a statement of an order or relation of phenomena that, so far as is known, is invariable under the given conditions." One of the things I have learned is that there are precious few orders or relations in the realm of human behavior that are invariable. It might be better to use the term "aphorism," which is defined as "a concise statement of a principle." I'm afraid, however, that "Lyle's Aphorisms" doesn't have the same ring to it, so "Lyle's Laws" it is.

The laws are not presented in any particular order and are not grouped in any way. They can be read sequentially or randomly, all in one sitting or days apart. They invite reflection and, as was seen in reader responses, lead people to remember similar lessons or experiences in their own past.

And so, dear reader, through the pages of this book your ship will come close to mine. I hope that as you cut across my wake you will pick up some of the spirits that I have accumulated during my voyage. They won't be evil spirits and, with any luck at all, they will improve your "fishing." Enjoy.

LEARNING

It is good to learn from your own mistakes. It is better to learn from the mistakes of others.

IOWA, WHICH BECAME A STATE IN **1846,** was surveyed in the early decades of the nineteenth century. The map of the country roads is not very exciting because, unless there is an interruption by a river or an unusually precipitous hill, it is a simple square grid with a road every mile. Every two miles, there was a spot for a one-room country school. It was in such a school, Columbia No. 2, that I received the first nine years of my formal education.

There were many advantages to those little schools, but curricular sophistication was not one of them. We did go beyond the three Rs of historic note, with elements of history, geography, civics, and even art and music. There was probably some science, but no explicit introduction to the scientific method. And there was 'rithmetic, but no algebra, and nothing approaching the rigor of hypotheses, theorems, and proofs.

Once in high school, however, we were introduced to the beauties of higher—but not much higher—mathematics. In algebra and then plane geometry, I remember being fascinated by the beauty and order of the proofs that we studied. How could anyone progress unerringly along these logical paths with all their theorems and lemmas and corollaries and be able, in the end, to write with such confidence, "QED"? It must have taken enormous prescience to map out this complex path through the mathematical maze. How could anyone do it?

It remained a mystery to me until I finally got to college where a mathematics professor explained that the route taken by the original prover involved many false paths and dead ends. The prover would try this, find out that it didn't work, then try that and find out that that didn't work and keep trying until, eureka! This works! Of course, the prover didn't tell us about all those mistakes, but he learned from them, even if all he learned was that a lot of things didn't work but that one of them did. As we learned the proof, we were learning from his success, but we were also learning from his mistakes.

Which brings me to Lyle's Law of Learning: *It is good to learn from your own mistakes. It is better to learn from the mistakes of others.*

Indeed, this could be considered the fundamental principle of Lyle's Laws. In essence, my primary goal in compiling the laws is to help my readers profit from my mistakes. There have been many, and I have learned a lot from them. Sometimes, like the time I severed the tendons in my thumb with a hacksaw, the lesson was a simple, "Don't do that again." (Don't worry. That won't be one of Lyle's Laws). More often, it has been a lifetime of errors that have led to a lesson that has—I hope—led to a change in my behavior and thus to a general principle that I believe is worth passing on.

So how does one learn from the mistakes of others? Mainly just by listening and paying attention. That is particularly effective when the "other" is still feeling the effects of the mistake. One day in the laboratory I heard one of my professors mumble something about just earning another oak leaf cluster on his stupidity medal. He had a thin film of an alloy deposited on a glass slide and had put a masking

material on it so he could etch a pattern into the alloy. Unfortunately, he had put the mask on the glass side instead of the metal side so,
when, he put the slide in acid, he etched away his entire sample. The
lesson I learned from his mistake influenced my laboratory technique for the next twenty years.

At first blush, this law could sound a bit cynical, as if I'm suggesting profiting from the misfortunes of others. Not so. If people
are doing interesting things, they are going to make mistakes. Their
misfortune would be if they did not reflect upon and thereby learn
from those mistakes. Our misfortune would be if they did not allow
us to profit as well.

Indeed, isn't education itself a process of learning from the mistakes of others? Of course, as with mathematical proofs, our teachers
don't tell us about all the false paths that have been followed, all the
things that people thought would work but didn't. They only present the principles that have been proved and that we can count on.
While we are learning about all the things that work, however, we
are also learning, by exclusion, about a lot of things that *didn't* work,
thereby learning from the mistakes of others.

There are at least two ways to violate this law. The first is to
observe the mistakes that other people are making but fail to learn
from them. We certainly see this in our everyday lives when people young and old observe the destructive results of drug use but
go ahead and use them anyway. The second—and probably more
common—is to completely neglect the experience of others, to not
even recognize that there are people out there who are having or
who have had experiences similar to our own. This may be a result
of ignorance, but it is more likely a sin of arrogance. Both are to be
avoided.

My advice to younger readers is to pay attention to your elders;
they have had more years to make mistakes and have usually learned
from them. They will be pleased—sometimes too pleased—to pass
their attained wisdom on to you. Sometimes it will seem that their

experience is a bit out of date, but remember that general principles really don't change.

To my older readers, I say, pay heed to the youngsters. They make mistakes and learn lessons too, and are more in touch with a culture that we elders don't always comprehend but from which we can often learn.

In conclusion, I would note that the Law of Learning is not unlike the Navy adage "Safety rules are written in blood," a reminder that the Navy's sometimes onerous regulations are based on the unfortunate experiences of the past. Learn from the mistakes of others.

HOW TO START

The most important step in solving a problem is defining it.

WITH APOLOGIES TO **JOYCE KILMER** ("I think that I shall never see / A poem lovely as a tree"), I have to point out that sometimes trees get kind of ugly. Certainly the one on our front lawn was no longer a thing of beauty, with its dead branches and the big rotten section on one side. It was time for that one to go, so, with the help of our daughter and her family (our son-in-law did the climbing and sawing while I provided tactical instructions from the ground), we cut it down, cut it up, and hauled it away. That was during spring break.

The absence of that scraggly tree helped the appearance of the place, but my victory was not complete; the *stump* remained. Over the weeks, it lost the color of new-cut wood, but still it was obtrusive. Kilmer described the stump and roots of a tree in beautiful words ("A tree whose hungry mouth is prest / Against the sweet earth's flowing breast"), but this stump was ugly and getting worse. Finally, after procrastinating as long as I could—and with the encouraging coolness of September—I dug it out.

Digging out a stump is an interesting process. The object of my attack was right in front of me (obviously). The problem I had to solve, however, was hidden from view beneath the soil and grass that looked so benign. So the first thing I did was to start digging that soil away. On the second or third spadeful, I hit a root—the first of many. The next two hours were spent locating, isolating, and excavating a total of seven major roots, ranging in size from six inches in diameter to almost a foot. Once all the roots were exposed, it was time to take a break because then I knew—I *really* knew—what had to be done to finish the job. From here on in, it was pretty simple: a little chopping, a little sawing, a little prying, and, voilà, the stump was no longer a stump but just a pile of wood chunks, ready to go to the dump.

Solving the problem of removing the stump was not unlike solving other problems that we face; once the problem was defined, once I knew where the roots were located and how big they were, the rest of the job was pretty straightforward. To generalize, I give you Lyle's Law of How to Start: *The most important step in solving a problem is defining it.*

Certainly this is a critical lesson for engineering students. I know that students are supposed to read the textbooks and work out the derivations, but the reality is that most students spend most of their study time working problems. They do this because they have learned that when they go in to take a test they will be asked to… work problems. This system has served us well over the years and will not likely change. But, oh, how much more efficient it would be if students learned to spend more up-front time figuring out what needed to be done before they started trying to do it.

For a number of years, early in my academic career, I taught the introductory electrical engineering course. One of my students seemed like a bright enough young man, but his performance on examinations was atrocious. He seemed destined for a career outside of engineering (in case you didn't recognize it, that's a euphemism for "flunking out"). His test work was so strange, though, that I decided to watch him as he took an exam. It was incredible. This was before our lexicon was invaded by computer jargon, but his actions could best be described as a memory dump. As soon as he received his test, he started writing down all the formulas he had memorized in preparation for this horrible hour. I went over and suggested that he slow down a bit, but as soon as I left, he was back to writing feverishly. When I corrected his test, sure enough, he received the low score once again.

In an effort to save this not-so-budding scholar, I called him in and discussed his approach to problem solving. What I found was someone who was so concerned about having enough time to do the work that he just *had* to start immediately. It was a classic case of "Ready! Fire! Aim!" but skipping the "Ready" step and never getting around to "Aim." So I made a deal with him. The next time he came in for an exam, I took away all of his pencils. I held them for the first ten minutes of the test time and then returned them so he could start writing. It was a tense time for him. He was clearly distraught for the first minute or two but then he settled down and actually read the problems and decided what he would do once he got his pencils back. To make a long story short, he passed that exam and, while he never became the top student in the class, he did pass the course and went on to graduate. I am convinced that he would never have done so had he not learned to define his problem before he started solving it.

This same principle applies as well to engineering problems of the nonacademic variety. If an electronic "box" is getting too hot, an engineer might logically start trying to improve the heat removal system. But think. The problem is a hot box. That might mean too much heat is being generated, not that too little is being removed. In solving a problem, it is at the very beginning—when the problem is being defined—that the problem solver must be skeptical. And it is there that you will enjoy the greatest return on investment of time. The old saying, "Don't jump to conclusions," modified by someone to "Don't jump to exclusions," is excellent advice.

Finally, Lyle's Law of How to Start is also very useful in solving the problems that arise in the process of living a life. I think that everyone, at some time in their life, feels that they don't have enough money. People solve that "problem" in many different ways: by trying to earn more, by engaging in some illicit money-making activities, or just by going from day to day in a blue funk. It would be better to spend some time thinking it through and defining the problem. Do I really *need* more money or do I just *want* more money? Is money the issue or is it the things that money can buy? If the latter, why do I want those things? Again, once the real problem is defined, solving it often becomes pretty straightforward.

Engineers are, in general, very logical and orderly people. One manifestation of this trait is an ongoing search for problem-solving algorithms that allow us to obtain solutions by performing a set of predefined steps. This law is not such an algorithm, but keeping it in mind when approaching a problem can save a lot of wasted effort.

WHINING

Curse the darkness—but not for long.

ave you ever noticed that many families develop a special language—words or phrases that make no sense to an outsider but convey a meaning to other family members? In our family, we ate "mitmy" because that was as close as our oldest daughter could come to saying "ice cream" when she was learning how to talk. When someone went unstable, they were said to "spaz out" or "go beseeched" (never mind where those came from). And my son and I still say that things are "simular" instead of "similar" because that was a pronunciational peculiarity of one of his teachers. We had dozens of such expressions and even today, when we are together, some of these words surface, much to the consternation—and mystification—of newcomers to the family.

Of course it is not just natural families that generate special languages. When I was deaning, the associate deans and I formed a very close-knit group and, in the manner of a family, shared some expressions that others found a little strange. For instance, when we were discussing an issue and it became apparent that it was outside our area of expertise, it would be identified as "an atomic-bomb problem." We didn't know much about atomic bombs, either.

Probably the most useful expression, however, would usually surface when we were faced with an unpleasant situation such as a

budget cut or insufficient laboratory space or a bad printing job. We
would grouse about the situation for a while and make all the usual
observations about how unfair it was and how others were being
treated better, and so on and so on. Eventually, however, someone
would invariably say, "Well, now that we have cursed the darkness…"
That was our signal to move on. The whining had been done but
now it was time to figure out what we were going to *do*. We learned
our lesson so many times that I think it is fitting and proper to generalize it as Lyle's Law of Whining: *Curse the darkness—but not for long.*

The complaining was not totally unproductive. It was a form
of catharsis, letting us vent our spleen and express our frustrations.
(As an aside, I love that "spleen-venting" expression. I know that
"spleen" means ill will and bad temper and other such stuff that we
need to vent or get rid of. On the other hand, the expression conjures up visions of a visceral organ exuding some nasty stuff. Much
more colorful.) The downside of darkness-cursing, though, is that it
tends to focus attention on what others are doing *to* us and not on
what we can do *for* ourselves.

Students often develop a penchant for darkness-cursing. I presume that this is because many of the unpleasant things that happen
to them are, indeed, associated with an action taken by someone
else. An example might be a pop quiz given by a teacher. *Curse* that
teacher who gave a quiz when I wasn't prepared! Or the student
might have to take a class at 8:00 a.m. when he really preferred an
11:00 a.m. section. *Curse* that person who schedules the classes! Or
the cafeteria food isn't as good as Mom's. *Curse* that food-service
cook! (A little Navy wisdom here: *never* curse the cook.)

Of course, the practice is not limited to students. When I received my bachelor's degree, I went to work for a large aerospace
company where I started out in what was known as the Evaluation
Department. The new engineers didn't have desks. Instead, we were
located at workbenches, where we got the not-so-subtle message
that we were there to work and to get our hands dirty doing it. At

the bench next to mine was a guy who griped from the time he arrived in the morning until the end of the workday, an event that he observed with great punctuality. While most engineers spent a few months to a year in Evaluation before moving on to design or another more advanced job, this guy had been there for about three years. And I presume he was cursing the darkness all that time. It never occurred to him that all of his cursing wasn't solving his problem. I have no idea whatever happened to him, but I know he was still there when I left.

So, once the darkness has been dealt with, what should you do? Well, the first thing to do is to change the focus. Instead of concentrating on the outside and analyzing what others are doing to you, look inward—at yourself or your organization—and see if you are contributing to the situation that has you so upset. The student who flunked the pop quiz has no reason to curse the teacher. The problem was not with the quizzer, but with the quizzee. And the engineer who couldn't get out of the Evaluation Department? I can't believe he was not bright enough nor well educated enough to be successful. He just needed to turn his attention from the outside—the people he perceived as giving him bad assignments and bad evaluations—back to himself. He needed to light a candle. Or an even bigger fire.

Of course it is quite possible that we are *not* the main part of the problem. It does happen that unpleasant situations are inflicted upon us by someone else and that we are now simply faced with a new fact of life. The student who didn't like the 8:00 a.m. class probably didn't do anything to deserve being sentenced to a semester of early rising. But that doesn't change the need to redirect the focus from the outside to the inside. The world is not going to change to alleviate your darkness. It is you who have to take some action to mitigate, to adapt to, or perhaps even to take advantage of the new situation.

The best thing to do, then, is to go ahead and curse the darkness—for a little while. It feels good and you need it. Once that is done, however, it is time, as the coach used to say, to pull up your socks and start solving the problem. Stop looking at the outside and start looking at yourself. If you are part of the problem, see what you can do to change. If you are not contributing to the problem, come to grips with the fact that it is you who will have to come up with a solution, and get started. Whatever the source of the problem, *you* are the source of the solution.

4

THE LABORATORY

See what is there, not what you wish for.

TO THE DAY SHE DIED, my wife's Aunt Esther was a work in progress. Born a few years before the Great War (for our younger readers, that's World War I), she was a nurse, a wife and mother, a pillar of her church and community, and, in her middle years, a somewhat aggressive driver. She drove the two-lane Iowa highways as if she were always in a bit of a hurry and she had no hesitation about pushing the speed limits just a little. As a matter of fact, it was insufficient hesitation that brought Aunt Esther a brush with the law.

Somewhere—I'm not sure just where—Esther failed to come to a "full and complete stop" at a stop sign. As one might expect, a minion of the law was nearby, observed the infraction, and hauled her before the court. The judge asked her for her account of the situation and Aunt Esther (I can still hear her telling this story) said, "Well, Your Honor, I approached the stop sign, slowed down and looked both ways. Since I couldn't see any traffic coming, I decided to just mosey on through."

The judge deliberated very briefly and then passed judgment. "I see. Well, Ma'am, the next time you come to an intersection and see a sign that says 'MOSEY,' you go ahead and mosey. But if it says 'STOP,' you stop. That will be thirty dollars and two points."

This little story is illustrative of Lyle's Law of the Laboratory: *See what is there, not what you wish for.* There was a stop sign there but Esther wished for a mosey sign, so that's what she saw. We can find ourselves doing the same thing in the laboratory.

A few years ago, some of us spent some time discussing and arguing about the fundamental objectives of engineering instructional laboratories. There was not always complete agreement on what those objectives are. I know of no such disagreement, however, about laboratories that are used for development or research. It has been said that we go to the laboratory to ask nature a question. Less poetically, we may say that we go to the laboratory to determine— by experiment —the behavior of the physical world, be it natural or man-made. But, oh, how easy it is to see what we want to see and not what is there.

I'm not talking here about those few greedy and unprincipled researchers who blatantly fake data in order to secure funding or enhance their reputation. They are clearly beneath contempt. Rather, I'm referring to those more complex situations where a certain amount of fuzziness and uncertainty exists and a reasonable person might see what he or she wishes to see. If you really wish for a result, be very skeptical if you get it.

One of my graduate students came to my office once to report on some very favorable results that he had achieved in the thin-film lab. The device that he was developing had performed very nearly as he—and I—had predicted. I looked at his data and, sure enough, the plotted points were scattered close to a straight line with a slope that was close to our predictions.

"Nice," I said. "How many times did you run this?"

"Five times," was the reply.

"With the same result every time?"

"Well, no. Only the last time."

So we had a little chat about skepticism in the laboratory and seeing only what you wish to see, and he went back to repeat the experiment. Naturally, his good results were not verified.

Was he faking data? I don't think so. He just wanted so much to see those results that he saw them.

Both engineering and science laboratories are susceptible to this kind of "wishful seeing," but I think we need to be especially vigilant in those devoted to the engineering development process. Invalid results from a science research laboratory certainly cause a lot of mischief—including significant financial loss—as people try unsuccessfully to repeat the published results or pursue unfruitful research paths suggested by the bogus data. In an engineering *research* laboratory, bad data can lead to wasted effort and funds as the results are used to start development of a product that is never going to work. It is in the engineering *development* process, however, where the greatest potential for problems occurs.

For one thing, in the development process we are usually in the laboratory to demonstrate that the product under development meets or exceeds a set of predetermined specifications. In those circumstances, what we "wish to see" is very well defined. That can make it easier to see than if we are just wishing for something vaguely defined as "good."

More importantly, however, unjustified results in the development laboratory can result in a product that cannot do what it is expected to do. That can be serious. People can be injured. Fortunes can be lost. People can die. If there is ever a place for skepticism in the laboratory, this is it.

As usual, a good law has a general applicability outside its narrow sphere of definition. Certainly that is true of the Law of the Laboratory. I don't think I know anyone whose life is so perfect that he or she does not sometimes wish to see something that is not there. It may be something as simple as a mosey sign. Or as complicated as the character of another person—someone we wish to trust or even love and spend the rest of our lives with. In both cases, it is far better to see what is there and not what we wish for.

And, of course, there is the matter of looking at ourselves. In "To a Louse," the great Scottish poet Robert Burns wrote,

> O wad some Power the giftie gie us
> To see oursels as ithers see us!
> It wad frae mony a blunder free us,
> An' foolish notion

Indeed what a "giftie" that would be. Assuming, of course, the objectivity of those others who are seeing us. In any event, it is useful to take the viewpoint of another person while trying to achieve our own objectivity as we examine ourselves, our behavior, and our motives. And seeing, we hope, what is there instead of what we wish for.

I am not very good at computer graphics but if I were, I would draw a stop sign with "STOP" replaced by "MOSEY" and then place across it a diagonal line, the universal sign for "don't." A copy of that on my desk would be a good reminder of Aunt Esther and her brush with the law, and also of my need to always remember to see what is there and not what I wish for.

THINKING

Think like a river

WHITEWATER RAFTING WAS NEVER high on my life list of things to do, but when my wife and I were invited to join our daughter and her family for a day on the New River of West Virginia, we were off to the rapids. This was not one of those adventure tours that puts a crew of 20 in an inflatable raft the size of a semitrailer to plunge down through waves that would get the serious attention of a destroyer escort. We rode in simple two-person inflatable kayaks and the most significant rapids were rated class III.

Not that it was a piece of cake. On one stretch of fast water, my granddaughter and I had an intimate and vigorous encounter with a submerged rock and found ourselves facing cross-river as we arrived at the big standing wave in the middle of the rapids. Bad position. The kayak flipped and we got a dunking. There were no injuries (except to my pride) and we drifted into the next area of calm water and got back into the boat.

Arriving in that section of flat water made me appreciate the fact that the New River changes its character from time to time as it follows the laws of physics that draw it ever downward. In some places, its waters converge and move rapidly forward. In others, they diverge and slow down and lap gently at the shores, which are broad and relatively placid. This convergent-divergent pattern, repeated throughout the river's course, led me to the discovery of Lyle's Law of Thinking: *Think like a river.*

What can a river teach us about thinking? First consider the rapids. When we are in problem-solving mode, our thinking is like the flow of water through the fast stretches—convergent and linear. The solution (i.e., *the* solution) is somewhere out ahead of us, and our job is to get there in the fastest way possible, with each step dictating, or at least suggesting, what the next step should be. We are in the rapids, moving with as much speed as we can muster, buffeted about a bit but drawn ever forward by the principles of problem solving and the desire to reach the solution as soon as possible.

But wait: The river doesn't stay in the rapids forever. After a period of fast, convergent, sometimes turbulent progress, it slows down, diverges, and eases along between the wider shores, considering, perhaps, how to proceed when it enters the next constriction. Indeed, rivers do sometimes find new paths out of these placid pools and stop following the channels previously pursued (witness the oxbow lakes in many river valleys). We need to do the same in our thinking. We need to pause occasionally in our pursuit of a solution to let our thinking spread out, to see if we have missed anything, to consider if it is really time to lock ourselves into this particular path or if there might instead be a better, more creative way to accomplish our ends.

Of course, we don't have the luxury of staying in this relaxed, reflective mode for very long. An engineer's job is to solve problems and that doesn't mean just reflecting on them or considering various alternatives. After an appropriate amount of divergent thinking, we have to enter the rapids again and speed things up and move once again toward a solution. The trick of thinking like a river is to have the discipline to hesitate from time to time and say, "Now that I know where I am going, am I sure that's where I want to go?" To slow down and allow the right side of our brain to take over for a while and be divergent and creative. To look for alternative solutions. To consider some of the consequences (environmental? social?) of the path we are pursuing. And then back to the rapids.

Of course, this metaphor eventually breaks down, as all metaphors must. Generally, as rivers approach their appointment with the sea, they slow down and spread out into a delta with no clear conclusion. Obviously, that is not a good model of thinking for engineers or anyone else. We dare not think like the Mississippi in its lower reaches. There are some good engineering rivers, however. Take the Niagara, for instance. It has its rapids and its cascades; it even has a major (how's that for understatement?) waterfall and at least one whirlpool. It also has some wide spots where it moves with relative calm. Finally, it enters Lake Ontario in a clearly defined channel, having resolved all the uncertainty of its passage—a great model of a good thought process.

This is probably as good a time as any to introduce the reader to Lyle's Law of Laws:

The better the law, the more general its applicability.

I use this law to judge the quality of proposed laws and decide which one to write about. By this criterion, the Law of Thinking is outstanding. It certainly does not apply just to engineers. Indeed, while these paragraphs have been directed to those of us of the convergent persuasion to remind us to broaden our thinking from time to time, the law can as easily be used as an admonition to the right-brained among us that, eventually, convergence is necessary so that action can be taken.

And the law can be useful in many facets of our lives. We make various decisions as we speed through our days and weeks and months and the quality of those decisions could undoubtedly be improved by thinking like a river. Take the time to ask whether your current decision is indeed the best, whether there are better alternatives and, indeed, *why* you have chosen that particular option.

Some say that a river is a living thing—even that it has a soul. I'm not sure I would go that far, but I do believe a river is a great model of how to think. If that means it is alive, so be it.

COURAGE

Be brave, but don't be fearless.

MY FAMILY HAD CHARTERED A PARTY BOAT in the Florida Keys, complete with captain and crew, food, drink, and snorkeling gear. Our snorkeling experience started in a shallow bay where we all got wet and practiced swimming around with our faces underwater, a decidedly unnatural activity. We then got under way and headed offshore to a spot in the Florida Straits, several miles from land. Here, the captain tied the boat to a mooring buoy and told us we could get back into our snorkeling gear. We were at the reef.

Well, it looked like just so much ocean to me, but he insisted the reef was just ahead of us. All we had to do was swim about 100 yards to the east and we would have some great snorkeling. We put on our masks and fins and milled about a bit and a few of us got into the water and headed eastward. As I was waiting my turn at the ladder, one of the family aquanauts came over and said to me, very quietly, "I'm just terrified." All I could say before I started to descend was, "To tell the truth, so am I."

Once in the water, I swam to the east (I guess) and sure enough, there was the reef, with fish of all shapes and colors and a rainbow of coral fans and horns and branches. I raised my head and looked about and there was the reluctant family member, right out with the rest of us—brave, to be sure, but still not without fear. Being brave allowed her to have an exhilarating experience in and on the water. Being afraid led her to do so with a healthy amount of caution.

Standing there on the deck, it would have been easy to say, "Oh, don't be afraid." At least it would have been easy if I had not had a measure of fear myself. But it would have been the wrong message, because fear, it turns out, is a great ally. Courage helps you succeed. Fear helps you stay alive. Let's take a look at Lyle's Law of Courage: *Be brave, but don't be fearless.*

I think there is a lot of misunderstanding—or at least linguistic imprecision—about being fearless. We have all heard reference to "our fearless leader." It is a great compliment to be called fearless and this appellation is intended to inspire confidence in the leader's ability to take us to wonderful places and to help us achieve great things. But do you really want a leader who knows no fear? When hiking through the mountains, might it not be best to have a leader who is afraid of falling off cliffs and hence helps us avoid that unpleasant eventuality? It might also be nice to have one who is afraid of being bitten by a rattlesnake and who will lead us around, rather than through, the rattlesnake den.

There is an adage among seafarers that says, "A sailor who is not afraid of the sea will soon be drowned." Good advice. You'd *better* be afraid of the "old gray widowmaker" because if you aren't, she will kill you. But does that mean you never go to sea? Of course not. There are two basic issues: is there reason to be afraid, and, if you are afraid, how do you deal with it?

First we have to recognize that fear is rational and reasonable only if there are valid reasons to be afraid. We can dream up all kinds of dangers if we walk in the dark, but if we shine some light on our surroundings, we can often see that the dangers are really not there. There is good reason to be afraid of the sea, but many of our fears are unfounded and, with a bit of examination, can be eliminated or at least diminished.

But what if your objective analysis leads to the conclusion that yes, there are good reasons to be afraid? What do you do about it? When Franklin D. Roosevelt assumed the presidency in the darkest days of the Great Depression, he said "The only thing we have to fear is fear itself." I don't think he was telling his people not to be afraid. He was telling them to cast out "nameless, unreasoning, unjustified terror" and then to deal with their very real and justified fears and to be courageous in the face of those fears. Indeed, the case can be made that one can only be brave if one is afraid. Bravery is not the absence or the abolition of fear, it is the ability to do the things you are afraid of doing or to go to the places you are afraid of going. To get anywhere, we need to summon the courage to act in spite of our fears.

So far, most of my examples have been of fear of physical dangers that we might face. But we also have fears that are not so physical in nature—those fears that we encounter in our professional and personal lives. They are no less real.

Consider, for instance, the fear of failure. Is this a rational fear? I wish I could say that it is not, but I can't. Whenever you begin a new undertaking, there is a very real possibility that it might not succeed. Any number of things can go wrong. Some will be under your control, some will not. Let's face it. You can fail. This is no time to be fearless. Your fear of failure should now inspire you to give the undertaking, whether it is a new job or a design project or a personal relationship, some very careful scrutiny. Do a thorough analysis to identify those things that can cause failure and then do what you can to minimize that eventuality. You may find that there is really not so much to be afraid of after all.

On the other hand, you may well find that there are lots of reasons to be frightened. Now is the time to summon your courage. To decide that even if your worst fears are realized, it won't be so bad. To decide that the potential rewards outweigh the possible costs. To decide that, no matter what, this is the right thing to do. In other words, to be brave, but not to be fearless.

James Norman Hall, coauthor of *Mutiny on the Bounty,* wrote this poem:

> FEAR
> The thing that numbs the heart is this:
> That men cannot devise
> Some scheme of life to banish fear
> That lurks in most men's eyes.

Dare I add a verse of my own?

> ADDENDUM TO FEAR
> Men's eyes could lose that lurking blight
> If we could banish fear,
> But taming fears that still exist
> Can also make eyes clear.

There. I have overcome my fear of being judged a lousy poet.

LYLE'S LAW OF

WHY

For everything there is a reason, but it may not be obvious.

WE HAD BEEN STUDYING SPANISH for a few days when our teacher led us through an exercise in which our task was to say that we liked or did not like various things. For instance, if I liked ice cream—which I do—I would say, "Me gusta helado." If I didn't like chocolate—which is, of course, ridiculous—I would say, "No me gusta chocolate." The teacher's primary goals were to drill us in the use of "no" as a negation and to help us expand our vocabulary of nouns.

One student, though, asked why we used "me" (pronounced "may"), the Spanish word for "me," instead of "yo," the Spanish word for "I." The teacher's response was, "Well, that's just the way it is." And we went on our merry way gusta-ing this and no-gusta-ing that.

But I was troubled. There may be a few things that come without explanations, but I can think of nothing that comes without a reason. Surely our teacher knew why the sentence was so structured and I wish he had taken the time to give us the reason. This and other similar experiences have led to the formulation of Lyle's Law of Why, the first part of which is: *For everything there is a reason.*

At first blush, this may seem like a trivial observation—at least to engineers and scientists. I am always surprised, though, at how many people are satisfied with the "well, that's just the way it is" response. Or, if they do believe there is a reason, either conclude that the reason is not knowable or, even worse, settle on a reason that is incorrect (although it may be logical).

One day when I was a college freshman, my English teacher started class with a pop quiz that consisted of a single question: What does "post hoc, ergo propter hoc" mean? Well, I hadn't studied Latin—nor had I read the day's assignment—or I would have known that it means "After this, therefore because of this." It is a classical fallacy of logic wherein one incorrectly concludes that because event A preceded event B, event B is a result of event A. Well, who would fall for that? Let's stay with Latin for a bit.

Recently I read an Associated Press article that described the comeback of Latin as a high school subject. While I think this is a good thing, I was a bit troubled by some of the statistical conclusions they used to tout the benefits of studying the language. They reported that in 2002, high school students who had studied Latin had a mean score of 559 on the verbal portion of the SAT while those who had studied French achieved 524, and Spanish, 501. Their conclusion was that studying Latin raises a student's verbal score. In my humble opinion (probably not as humble as it should be) the data do not support that conclusion. I would suggest there is likely a significant student self-selection process that results in Latin classes being populated by people who are already verbally adept. Post hoc, ergo propter hoc? Not necessarily.

This leads me to round out Lyle's Law of Why with an addition: *For everything there is a reason, but it may not be obvious.* Clearly, you need to look for the reason—and then look again.

For the most part, I think engineers believe in and live by the first part of the Law of Why, at least in their engineering work. Causality is more than just a creed for engineers; it is a principle that underlies all that we do. The second part of the law, however, can sometimes be forgotten or ignored. This can happen in two ways.

First, in solving a problem that has cropped up (the process we call troubleshooting) the first thing we need to do is define the problem. The second is to look for the cause of the problem or the reason that it has occurred. Once that reason has been determined, we can then proceed to fix the problem. But don't jump to conclusions on that reason. The real reason, or the root cause, may not be obvious.

"The dining room lights won't work."

"Aha! The circuit breaker is tripped. I'll reset it."

"Okay, they work now."

Never mind the lamp with the intermittent short circuit. Uh oh!

Another place where we sometimes ignore the second part of the law is in the design process. If the product fails, there will be a reason, *but it may not be obvious*—especially in the design phase. It is probably impossible to identify all of the variables that will affect the product we are designing but we need to do the best we can.

Some years ago—quite some years, actually—I found myself teaching a course in power systems. This is not exactly my field of expertise, but the students and I made the most of it and we all learned a lot. Since I didn't have a great deal of experience in the field, I tended to give quite a few open-ended problems. One was to do the preliminary design for a transmission line to bring power to a village high in the Andes. I was very proud of my students because, with a minimum of prompting, they recognized that altitude influences the corona effect, that jungles grow rapidly, and that the climate can influence the choice of transmission towers. One group even asked if we were sure those people *wanted* electricity. They looked for the reasons that may not be obvious.

So never settle for, "That's just the way it is." And never believe that the reason is always obvious.

Incidentally, when I learned a little more Spanish, I understood the reason for using "me" instead of "yo" in that language drill. When I said, "Me gusta helado," I was not saying, "I like ice cream." I was saying, "Ice cream pleases me," which explains the usage and carries the same meaning. How much more satisfied we would have been had the teacher given us that explanation? Why didn't he do it? I don't know. For everything there is a reason. But it may not be obvious.

BECOMING

What you are becoming is as important as what you are doing.

O UR INSTRUCTOR, STAN, WAS A NERVOUS SORT. And justifiably so. He was engaged in teaching five neophyte sailors how to maneuver a thirty-seven-foot sailboat around the waters of Long Island Sound and hoping to get all of us—and the boat—safely ashore again. As any boater knows, one of the most critical and difficult parts of boating is docking. It was in drilling us on this operation that poor Stan had an ongoing flirtation with a heart attack. He had more ways of using the word "slow" in a sentence than I ever imagined possible.

Of course, Stan's admonition to approach the pier slowly was good advice, but it didn't completely cover the operation. The limit of "slow" is "stop," and if the boat stops too far from the pier, the skipper's goal has not been achieved. The perfect landing is when the boat's velocity and its distance from the pier reach zero at exactly the same time. To achieve this, the skipper must be aware of not only the boat's position, but also its velocity. Control systems engineers call this velocity input "rate feedback" and recognize its essential role in stabilizing a system.

Velocity—or rate of change—is important in many different

kinds of systems. While you need to know where you are, it is often even more essential to know how fast you are going and in which direction you are heading. When applied to human beings, this becomes Lyle's Law of Becoming: *What you are becoming is as important as what you are doing.*

In your working life, you are expected to do a job, no matter whether you are self-employed or work for a large corporation. To prepare yourself for doing that job, you have invested a lot of years, considerable sums of money and an enormous amount of energy gaining an education and the required experience. As a result, you probably now find yourself in a very good position. But look at that word, "position." It is where you are. It is what you are doing. It says nothing, however, about what you are becoming. This requires some separate attention.

Such an observation could engender an admonition to seek a graduate degree or to be involved in continuing education. While these are very important career development activities, they may be both more than you need and less than you need. There are lots of good jobs that don't require an advanced degree and it may be hard to find another short course on just the right topic. Taking care of what you are becoming may require some kind of continuing education, but it also demands something more. Just what it demands will vary from person to person, but there is a simple way to check your velocity.

Start by asking yourself this simple question, "What can I do today that I couldn't do a year ago?" If the answer is "nothing," that would suggest that you are becoming just what you are today. And that's probably not good enough. Your speed equals zero.

If, on the other hand, you can list some new skills, new capabilities, new relationships, you do have some speed. Now you can check your direction (the other component of velocity) by moving on to the second question, "Does anyone care?" or, more pointedly, "Will anyone pay me to use my new bag of tricks?" And then, "Will I *like* doing that?" If your answers to both are positive, hey, you're on your way! If not, you need to take another look at what you are becoming. You may need some more education, or a job rotation, or to attend some conferences, or to get involved in your professional society or…what? Whatever it takes to become what *you* want to become.

Don't confuse this with goal setting. I know that a lot of career gurus advise you to set concrete goals and then pursue them. I won't quarrel with that technique, but I know a lot of highly successful people who didn't have any specific goals, who instead concentrated on just doing a great job, gaining a lot of diverse experience, and seeing what opportunities popped up. In other words, they didn't worry about where they were going but they did watch what they were becoming.

While I have been talking about individuals and their professional careers, this law applies just as well to managers. Yes, as a manager, you have a job to do and if you don't get it done, you won't be a manager for long. But while you and your group are doing the job, spend a little time thinking about what you are collectively becoming. If you are becoming the group that you already are, that's not good enough, because the jobs you are going to be asked to do next year are not the jobs you are doing now. You will have to upgrade your computers, get more advanced software and maybe a new set of instruments for the lab. And while you are at it, don't forget to upgrade your people and, of course, yourself as well. Pay attention to what your group is becoming, individually and collectively.

Of course, we are not just workers and managers. Engineers are people (the opinions of some *Dilbert* readers notwithstanding). In our personal lives, too, we are not only being; we are in the process of becoming. Our actions say something about what we are, but they also help determine what we will be. Take food, for instance—a perfect example of how our behavior determines what we are becoming. I know if I didn't limit (I didn't say eliminate) my consumption of cookies and pies, I would become about 220 pounds. I really want to become 210, partly because a weight of "fourteen stone" sounds so cool but mostly because I would be healthier. I'll get there.

Other behaviors have their impact. Giving, even a little bit, will turn you into a giving person. Constant complaining will turn you into a gloomy person—not to mention a pariah. Saving and wise investing will turn you into a millionaire. Spending more than you earn will turn you into a debtor. Being friendly will make you a friend. Voting will make you responsible. Lying will make you a liar. Well, you get the idea. What you are doing (saving, lying) is important. What you are becoming (a millionaire, a liar) may be even more so.

Once again, I will quote Lyle's Law of Laws: *The better the law, the more general its applicability*. Based on that, the Law of Becoming—*What you are becoming is as important as what you are doing*—is a good law.

LANGUAGE

Call everything by its right name.

FOR A PERSON WHO DOESN'T SAIL, the deck of a sailboat is pretty confusing. Some of the things on and about the deck are unfamiliar and others, such as ropes, are there in such profusion that it is hard to know just which one does what or why. The old square-rigged sailing ships of the eighteenth and nineteenth centuries were even more complex. While a modern sloop has eight or ten ropes, the older square-riggers had many dozens.

Wait. Time out. The sailors among you are asking, "What's with Lyle? Boats don't have 'ropes.' They have lines." Well, you're right, but I thought I'd bring the non-sailors along gently. Let's clarify it, though: the tradition is that a rope is a rope as long as it is on a spool in a marine store. As soon as it is cut to length, taken aboard a boat and assigned a function, it becomes a line. So, there is your useful (?) nautical fact of the day.

Now back to the square-riggers. What wonderful, beautiful machines they were! They were driven through the sea by dozens of sails: fores and mains and mizzens, courses, topsails, topgallants, royals, skysails, spritsails, jibs, staysails, and even one called a spanker. These sails have heads, tacks, and clews, a leech, a foot, and a luff. They are controlled by lines called halyards, braces, sheets, buntlines, outhauls, downhauls, reef points, and many others.

Sound confusing? You bet. And the situation would have been impossible if each one of those lines didn't have a name and a place: the "starboard foretopsail brace," the "main yard sprit outhaul," and so on. The only way the ship could operate was if every sailor "knew the ropes" and could find the right line in the dark of night on a rolling deck. And you couldn't have the boatswain giving the order, "Okay, you guys. Pull in the whatchamacallit." No, to make it work, everyone had to follow the principles of Lyle's Law of Language: *Call everything by its right name.*

There are several things that can lead to the violation of this law; one is plain old-fashioned ignorance. A twenty-first-century engineer can be forgiven if she doesn't know the name of the line that controls the angle of the third sail from the bottom on the main mast of a square rigger. She must, however, know the terminology of her field.

There are various educational taxonomies that classify the kinds of learning according to the level of intellectual challenge involved. The lowest or simplest category is generally memorization or the learning of definitions. Since it is not considered very challenging, many professors think they should devote very little attention to teaching terminology. I disagree. Students must learn the terms that define their field and not only be able to provide a rote definition but also understand the fundamental meaning and usage of the word. Students and practicing engineers can do themselves a big favor by keeping a list of new words and—surprisingly—new meanings for old words. You can't call things by their right name if you don't know the name.

While knowing the names of things aids in communication, it also helps the thinking process. Indeed, a case can be made that you can't think about something unless you have a name for it. Try it some time. Can you think about something that doesn't have a name?

A corollary to Lyle's Law of Language is: *Eschew euphemisms.* A euphemism is a word or phrase that is used in place of another word or phrase and is generally used to soften or obscure the message being delivered. "I am experiencing financial difficulties," is less harsh than "I have just declared bankruptcy." Both statements are true, but the latter is more precise. It calls the thing by its right name.

In the spirit of full disclosure, I will point out that this law is taken directly from the writings of Confucius. One of the many things he said is, "The beginning of wisdom is to call things by their right names." When I was a dean, this phrase was used from time to time in our staff meetings. For instance, we sometimes had to deal with cases of students "copying," or "collaborating," or "lifting answers." The beginning of wisdom was to say they were *cheating*, a thoroughly reprehensible activity that should disqualify the perpetrator from entering the engineering profession. With the malefaction properly named we could more effectively—if not more easily—deal with it.

I have to admit that using the right name is not always as easy as it sounds because rightness, like beauty, is in the eye of the beholder. Or more precisely, the brain of the hearer. President Bush learned this when he referred to the invasion of Iraq as a "great crusade." The word "crusade" has come to mean any major campaign undertaken with great zeal and noble motives. Many people in the Arab world, however, remember the original meaning of the word: a military expedition by Christians to take control of the Holy Land ("crusade" is derived from the Latin *cruc*, which means "cross"). I'm not sure the president ever started calling the invasion by its right name, but he wisely stopped calling it a crusade.

I guess this is as good a time as any to get on one of my favorite soapboxes and address the "comprise" situation. Calling things by their right names also means using the correct word and using the word correctly. "Comprise" is a very useful word, but it has been misused unmercifully. I first got serious about the word some forty years ago when I was reading patents and read something like, "The circuit comprises two AND gates and two OR gates." My first reaction was to think they used the word incorrectly. My second was to realize this would not happen in a patent. Those folks have to be precise. I then studied up on the word and determined that, lo and behold, *I* had been misusing the word. Blush.

"Comprise," it turns out, means almost the same thing as "include," except that it is comprehensive. That is, to comprise is to include everything. As an example, a football team comprises eleven players and includes two tackles. An orchestra may comprise seventeen instruments and include five violins.

Somewhere in the past, however, something went astray and someone started using "comprised" when he meant "composed" and the unfortunate phrase "comprised of" was born. Going back to the definition, this is comparable to saying, "included of." While "comprised of" is regrettably common, most people consider it incorrect. It is to be avoided.

So. Call things by their right names. Eschew euphemisms. Use the correct word. Use the word correctly. Cast off the bow line. Haul in the jib sheet. And I will go splice the main brace. Splice the main brace? That's not really a euphemism, just another bit of nautical terminology. Old sailors may know what it means.

10

EXPECTATIONS

Model success. Expect the best.

Some years ago, I was asked to serve as master of ceremonies at my high school reunion. One of the features of the program was a letter from the man who had been our football coach when we were seniors, and it fell to me to read the letter to the assembled classmates. Coach Sam either had a great memory or a good scrapbook because he listed the starting lineup of the 1952 season. So, of course, I read it:

> Center, Kucera
> Quarterback, Doran
> Right end, Porter
> Bench, Feisel

Well, he didn't really say that about Feisel, but I thought I should insert it before one of my less gracious classmates asked where Lyle fit in the lineup. The fact is that I was a less-than-outstanding football player. I believe that several factors contributed to this. For one thing, I was a farm boy and didn't get started in football until I was in high school. For another, I had an aversion to avoidable bodily pain, a condition that exists to this day. Mostly, however, I think my lackluster performance was due to the fact that I never did visualize myself as a great football player. As I ran the plays in my head, I didn't see myself as the hero of the game or even an important contributor. I think it would have helped if I had been able to run some scenarios that took me through the plays and put *me* in the starring position. It would have been better had I *expected* to be good. This principle is summed up in the two-part Lyle's Law of Expectations: *Model success. Expect the best.*

Airplane pilots of the amateur variety do this. A pilot friend of mine once told me that he shot landings in his head, visualizing the ground coming up to meet him, the end of the runway passing under the plane, and finally, the plane leveling out just before the wheels touched the surface. As these events occurred, he was busy visualizing his actions, managing pitch, roll, and yaw, adjusting the throttle, and monitoring the instruments as the aircraft responded to his inputs. And of course, every landing was perfect. He did the drill over and over and, by modeling success, came to *expect* that he would do it just right and the people on the side of the field would cheer.

Salespeople follow this law. A successful salesperson will visualize the introduction, the approach, the rationale, the objection, the rebuttal, the pitch, and, finally, the sale. In their mental model, they *always* make the sale. That is, they expect the best to happen. And they *act* as if the best is going to happen. You won't hear a good sales person start a sales call by saying, "I don't suppose you'd care to buy…"

This law is useful in the design process. A designer can start out by visualizing a product that is able to do all the things one could possibly desire in such a product. Usually, then, one good visualization leads to another. The thumbwheel on my cell phone rolls up and down, pushes in, pushes forward, and pushes backward—five degrees of freedom. I'm pretty sure that in the first visualization, it only had one. But hey! Let's dream a little.

As the design process continues, the designer can visualize individual circuits, mechanisms, processes, etc., not from the standpoint of how they work, but how they perform. With this input, what is the ideal output? Visualize the thing working in the best possible way and then expect that you will be able to build it to provide that kind of performance. *Model success. Expect the best.* And then go to work.

In management, this law is absolutely essential. I am very fortunate to be a member of the board of directors of a company founded by one of my former students. What a pleasure it is to be around Bob and his colleagues. They share a marvelous vision of this company as a world leader in its field, with a mission to make our world safer, a high growth rate, the confidence of its customers, and the support of its investors. Bob, in particular, has in his head a model of what the company can achieve and the expectation that it will do so. And it will.

In visualizing your model, don't make it utopian. As a matter of fact, one of the benefits of modeling success is doing so while integrating adversity into the model. My pilot friend would visualize landing in a crosswind or a rainstorm or at night. He would imagine engine failure at 500 feet. But always he modeled a successful landing. And all the people would cheer.

So as you create scenarios, put in the bad things as well as the good. Then model your way out of the negative situation and toward a successful conclusion. Imagine a part failing in one of your designs. Then visualize yourself fixing the part and the problem and preventing it from happening again. Imagine yourself in front of a hostile audience explaining and defending an action they don't like. Conjure up their objections and model your response. See yourself doing it calmly and logically and with sensitivity to the feelings of the audience and finally winning their support. And, of course, when you have done the modeling of success go do the real thing with an expectation that the best will happen.

Of course this law, like most, has its limitations. I can visualize myself as a great singing basketball player 'til the Queen of England comes to call, but I'll still sing like Michael Jordan and shoot baskets like Michael Jackson. (Note in passing: I almost wrote, "'til the cows come home," but that isn't long enough. If you know cows, you'll know they do come home. Twice a day.) Your model has to be realistic but positive. No matter what the adversity, you need to see yourself as succeeding and then expect to experience that success.

The Law of Expectations applies to football, to flying, to sailing, to design, to management, and to many other things. Above all, it applies to life. Certainly you should not spend all your time modeling and visualizing. You have to spend most of your time doing, whether you are doing work or doing play or doing relationships. But some time spent modeling successful work or play or relationships will help you build confidence. And the expectation of success will help you to be successful. So construct in your mind a model of yourself being a success in your profession, enjoying and excelling in your sports and leisure activities, and being a caring and supportive person to your friends and family and even to people you don't know. Then expect that you will truly *be* a success in doing all those things. Chances are that you will.

LYLE'S LAW OF

HEROES

Emulate the best. Respect the rest.

I T WAS A COURSE THAT PROMISED TO BE DIFFERENT. The exact
title is lost in the fog of the intervening forty-plus years but I do
remember that the subject was engineering analysis, the application of some new (to us) techniques for solving engineering problems. And not just problems of the electrical variety, even though it
was in the EE curriculum and was taught by a professor of electrical
engineering. On that first day of the quarter, that's about all we EE
seniors knew about the course.

The perplexing practice of students coming into class late was
still two or three decades in the future, so we were all assembled
on time. We were probably talking about the latest job offers and
the interview trips we were taking, when, precisely as the bell rang,
the professor walked into the room. I didn't know him. Whether he
was new to the department or whether I simply had never had any
contact with him, he was new to me. He walked to the front of the
room, wrote his name on the blackboard, went to the teacher's desk
and…sat down. Sat down? That's not the way professors behave. They
stand, preferably on an elevated platform so everyone can see them
and so they can establish their authority—both intellectual and hier-
archical—over the students. Well, not this professor. He sat, steepled
his hands in front of him, and proceeded to tell us what we would be

doing for the quarter. And he did it with such dignity, such authority, that he immediately gained the respect and admiration of the entire class. Throughout the quarter, he continued this mode of lecture, going to the blackboard occasionally to illustrate something that required a visual expression, but otherwise speaking to us from his chair.

As the term progressed and as I listened to and learned from this professor, he was added to my small pantheon of heroes, that illustrious (to me) group of men and women who have taught me so much about how to live. In learning from their examples, I was applying—long before it was formulated—Lyle's Law of Heroes: *Emulate the best.*

Emulate. My dictionary doesn't support my position, but I don't consider "emulate" and "imitate" to be synonymous. My understanding is that to imitate is to try to be the same as. To emulate is to try to achieve the same result but to do so by using different processes. Imitation may be the sincerest form of flattery, but I think emulation shows the greatest integrity and, to the emulatee, is the greater honor.

Why is this distinction important? We need to make it so that Lyle's Law of Heroes does not deteriorate into an admonition to engage in hero worship. The goal is not to be identical to our heroes. It is to look at them carefully, to determine what characteristics make them heroes and then to find our own way to develop those same characteristics in ourselves.

The tyranny of the dictionary must be dealt with in another issue as well—that of gender. My dictionary defines hero as being of the masculine gender. Not in Lyle's Law. Lots of my heroes are women, starting with Enid Edna Geiger, the one-room country schoolteacher who taught me for all nine years of my primary education. I realized after I had matured a bit that this woman loved teaching more than anyone I have known. She also managed to love her dozen or so charges, even though none of them was always—and some of them were rarely—lovable. What a hero.

I suspect there are people who have no heroes. I pity them. Perhaps they are good—or even great—people in their own right, but I wonder if they might not be better or greater if they recognized the existence of better and greater people than themselves and tried to learn from them. And I also pity them because if they don't look up to anyone else, they can look up only to themselves. It must be a dismal prospect.

Very often, a law suggests a corollary that seems so important that I attach it to the original. Such is the case with the Law of Heroes. Obviously, not everyone is a hero to you or to me. But I suspect that almost everyone is a hero to someone. As such, they deserve our respect, if not our admiration or even our approval. So the corollary is, *Respect the rest*. The point here is to avoid getting so wrapped up in your heroes that you reject anyone who is not one of them.

Our mothers told us if you can't say anything good about someone, don't say anything at all. We all know some people for whom this rule dictates a silence that makes the whisper of a flea sound like a brass band. But is it possible there is some good in everyone? Surely there is.

Finally, there is the "feet of clay" issue. No one is perfect, including our heroes. I have long been an admirer of Thomas Jefferson. There are, however, a number of things about Jefferson that are not so admirable. He owned slaves throughout his life, his protestations about the evils of slavery notwithstanding. He was something of a spendthrift, often in debt and rarely completely solvent. While he warned of the evils of having an aristocracy, he enjoyed the privileges of the landed gentleman. There are others. But he also had some heroic characteristics. He was a brilliant inventor. He wrote beautifully and prolifically. And his concept of the "natural aristocracy of virtue and talents" is an inspiration to everyone who believes that great things can come from those with humble beginnings. Those parts of Jefferson I can continue to admire, even as I criticize his flaws.

While we who are relatively long in the tooth have more leisure in which to contemplate this issue of heroes, it is the younger reader for whom the contemplation is most important. Budding engineers will be influenced by the heroes they choose, both in their personal lives and in their approach to the practice of our profession. Those heroes must be chosen with care.

Efficiently applied, Lyle's Law of Heroes—*Emulate the best. Respect the rest.*—is a classical positive feedback loop. One's character influences the selection of one's heroes. The members of one's heroic pantheon influence one's character. We are shaped in many ways by the people we admire. Who are your heroes?

UNIQUENESS

There are no ordinary people.

I T WAS A MOST MEMORABLE DAY. After ten days at sea, our freighter had completed the passage from San Francisco to Japan and was going slowly into the port of Yokohama. It was my family's first time in Japan; indeed it would be our first time on land in the eastern hemisphere. And on television we could see a very grainy picture of Neil Armstrong taking the first steps on the moon. The date was July 21, 1969, and it was the day of a giant step for mankind, and also a pretty big jump for the Feisels.

In our first exploration of Yokohama that afternoon, we toured the Silk Center, walked the streets looking at the small shops and also visited a large department store. It was a Monday, so all the young people were at work, leaving the department store populated by Japanese retirees and five Americans. Looking out over the crowd we saw a sea of uniformly black hair on people who were all of roughly the same height and, if you didn't look very closely, very much alike in appearance. I felt that we were seeing Joe and Jane Japanese multiplied by several hundred and all very much alike and very ordinary, at least in Japan.

As the years have passed, however, I have thought of that day in the department store many times and have come to realize that the hundreds of people we saw there were not really that ordinary or that much alike. They all had different parents, different spouses (or none at all), different children (or none at all), and different experiences (where "none at all" is not an option). Indeed, extending this thinking to ever-broader groups of people inspired Lyle's Law of Uniqueness: *There are no ordinary people.*

Biologically, it is quite clear that everyone is unique, right from birth. We are told that the human genome is a string of some 70 billion pairs of protein molecules arranged in different sequences. Of course, as I understand it, many parts of the sequence are more or less fixed, but there is still a lot of room for a vast number of combinations and permutations that determine individual characteristics. If "ordinary" implies pretty much the same as everyone else, biologically, we are not.

It is what happens *after* birth, however, that really makes each individual unique. We are all shaped by our experiences, by the people we have known, the things we have learned, and the places we have been. We have had different joys and different hurts, different successes and different failures, different ups and different downs. How, then, can anyone be ordinary?

As you advance in your education and in your profession, it is easy to forget this law. After all, if you are gaining all this knowledge, importance, and responsibility, surely you are being set further and further apart from the "ordinary" people around you. Well, maybe you are becoming more different, but you are not becoming less ordinary. You can't be less purple if no one is purple.

When I completed my bachelor's degree, I went to work for the aeronautical division of a large company. I was a bit surprised when I was shown to my workplace and found that it was not a cubicle or even a desk in a room with other engineers. It was a workbench, complete with test equipment and power outlets—and no telephone. Then I met the guy at the next bench, and, lo and behold, he wasn't even an engineer! He didn't even have a degree! He was an ordinary technician. Well, that may have been the beginning of the Law of Uniqueness, because engineer or not, that guy taught me a lot. He was not ordinary in any sense of the word and, as far as experience was concerned, he made me feel pretty ordinary.

Of course, he still had a lot to learn, too. Such as the old carpenter's law, "Measure twice, cut once." One day he was connecting 28-volt power to a flight control system when he failed to check the polarity and hooked it up backward—positive where it should have been negative, and vice versa. You can usually get by with that if you are using vacuum tubes, but transistors are very unforgiving. He had his fifteen minutes of fame but I wouldn't say that he enjoyed them.

For the past decade or so, engineering educators in the United States and many other countries have been placing a lot of emphasis on working in interdisciplinary teams. The reason usually given for this emphasis is that today's engineering problems are so large and complex that a team approach is required. I'm not sure they are that much more complex than the Manhattan Project or the Apollo Program or the 747 but I still think it is a good idea to stress teaming. If a team is to be successful, it can't be made up of ordinary people, i.e., all alike. The people need to be unique, each bringing his or her individual talents and perspective to the common effort. Fortunately, *there are no ordinary people.* Everyone can contribute something to the success of a team. It is the team's challenge to determine what that might be for each member.

In our private lives—the things we do outside of work—we also need to remember this law as we deal with the myriad people with whom we interact. The convenience store clerk is not an ordinary person. The paper carrier is not an ordinary person. The woman who keeps everything she owns in a shopping cart is not an ordinary person. I don't mean that we have to take responsibility for them. They do, however, deserve to be respected and, more importantly, not ignored.

There are two ways in which failing to remember this law can be harmful. The first—and probably most common—is to consider yourself to be special and almost everyone else to be ordinary. Probably even more destructive, however, is the converse of that failing—to consider yourself to be ordinary. Remember, *there are no ordinary people.* Look for your own uniqueness. What is different about you? Are these differences good? If so, develop them. If not, eliminate or suppress them. Above all, have confidence in your extraordinariness.

Of course, uniqueness is not always good. The worst characters in history were unique. Fortunately. But all in all, it is this astonishing breadth of humanity that makes our lives interesting and, indeed, productive and successful.

13

PALATABILITY

You don't have to like it.
You do have to deal with it graciously.

AMONG THE MANY CHARACTERISTICS that distinguish one culture from another, one of the most salient is food. Germany's sauerkraut is another country's rotten cabbage. Scotland's haggis is another country's discarded organ meat stuffed into a sheep's stomach. Mexico's gusanos asados are another country's yucky worms. In the United States, regional differences fall into the same category: grits, for instance, are not universally relished. And then there is lutefisk.

When we moved to Taiwan some 35 years ago, our children were all in elementary school and we knew that there could be some gastronomic challenges ahead. We talked it over and agreed on the following basic principle: You don't have to eat it. Just don't make gagging noises. Push it around on your plate a bit and wait for something more to your liking to come along. It seemed to work. None of the children starved, their psyches did not appear to be unduly damaged, and I don't think we offended any of our hosts. Puzzled, perhaps, but not offended. And so we maneuvered our way through the sea slugs, kidneys, eels, and fish balls, which I (usually) enjoyed and various other members of the family pushed around on their plates.

The principle worked for us in the food arena and I think it can be extended to other areas of the human experience. Restated as Lyle's Law of Palatability: *You don't have to like it. You do have to deal with it graciously.*

What, you ask—and well you might—is "it"? Clearly there are many "its" and I will attempt to deal with a couple.

First—and probably the most difficult—is people. Substitute "people" for "it" in the Law of Palatability and we have created quite a challenge. Some people are just not very palatable, due to their appearance, their behavior, or even their belief system. But we have to deal with them. Avoiding gagging noises is not enough. We at least have to push them around on the plate. Graciously.

Today, in industry and in education, much is made of the process of teaming. Clearly (Please forgive me when I do that. I always hated it when the math professor said "clearly" when it was not at all clear to me.), a team cannot be effective unless the team members are co-operating with each other. I think it is too much to ask, however, to expect that every team member will *like* all the other members. But everyone must deal with all the others. Graciously.

I knew a dean who had two department heads who just couldn't stand each other; they were apparently incompatible and, to each other, unpalatable. Their mutual dislike was hindering the organization because they refused to cooperate, and their departments, the school, and the students were not being well served. The dean called them in and laid down the law. You don't have to love each other but you do have to act as if you do—which is a pretty good interpretation of the Law of Palatability.

It is possible, of course, for a person to go beyond the merely unpalatable to being truly contemptible. In such a case, the law still can be applied, but in a somewhat different manner; gagging noises should still be avoided, but the situation needs to be met head-on. If, for instance, in the situation just described, Department Head A was involved in unethical or illegal activities and that is why Department

Head B didn't like him, then B has an obligation to deal with the situation. Graciously. Although I admit that being gracious in such situations is not easy. This is the most difficult part of the honor code or various codes of ethics. "I will not lie, cheat, or steal," is relatively easy; "nor tolerate those who do" is not.

Another interesting "it" is "the situation." The law restated: *You don't have to like the situation. You do have to deal with it graciously.* School, for instance. Ever have a bad professor? Unfortunately, most of us have. Obviously it is a situation that we don't like, but there is nothing to be gained by making gagging noises. I have had students tell me they were not attending class because the professor was so bad and of course they weren't studying the material either and, furthermore, they weren't passing the tests. Hoo boy! Deans will do their best to straighten out the bad professor but students need to deal with their own situation. (See also Lyle's Law no. 3, Whining. Whining and gagging are not the same but they do have the same antidote.)

One place where the Law of Palatability must be applied judiciously is in our jobs. Ideally, everyone loves his or her job and looks forward to going to work every morning, including Monday. Practically, that is not the case. Rather, even if you do love your job, there will likely be days when you would rather be doing something else. So you don't necessarily like it. But you did sign up for it so it is yours to deal with.

But I said that the law must be dealt with judiciously. Given that very few jobs are a joy all the time, neither should a job *never* be fun and satisfying. If that should ever happen to you, it is probably time to suspend the law and decide that the way you are going to deal with the situation you don't like is to get out of it. One can only push the food around on the plate for so long. Eventually, it is time to get up and leave the table. Life is too short to spend it in unrewarding pursuits.

As I look back on the original formulation of the Law of Palatability, I recognize that it came about primarily due to our own self-interest; we didn't want to be embarrassed by a kid pitching a fit because someone offered them a serving of sea slug. There were secondary motivations, however, that were much less self-serving and that I hope will generally guide the application of the law. One was the desire to protect our children from having to do things they really found disagreeable. The other was to keep our gracious hosts from being embarrassed because they had not thought to serve wieners and macaroni and cheese. The best laws, while guiding our own behavior, are motivated by their benefit to other people. Let it be so with this one.

LUCK

The more you learn, the luckier you get.

L AS VEGAS HAS CHANGED A *LOT* IN THE PAST FIFTY YEARS. It has grown like a pimple during prom week as billions of dollars have been spent on casinos and hotels—building, then razing and building again in an accelerating cycle. There has, however, been one fixed star in the Nevada firmament: the presence of the ubiquitous—and, some say, iniquitous—slot machine. The one-armed bandits may have changed from mechanical to electronic in their operation, but their function is still the same—to entertain the users while simultaneously separating them from their money.

Some thirty years ago, our family stopped in Las Vegas en route to California where I was going to work for the summer. We parked our camper at Lake Mead and drove in to the city to have dinner and do a little sightseeing. There were, of course, slot machines in the restaurant where we ate, so I decided it would be a good time to demonstrate to our kids, who were in their early teens, the futility of gambling. I popped a nickel into the machine, pulled the handle and…heard the "ching, ching, ching" of 13 nickels dropping into the payout hopper. Well, there went *that* lesson. Lady Luck had conspired against me and *she* won.

But I'm afraid there really is no lady named Luck. At least I don't think so. We are, however, affected by events that are apparently random in nature. People who are positively affected are deemed to be lucky and those who are negatively affected are identified as poor, unlucky so-and-so's. But are all those events really random? Or is it possible that you are making some of your own luck? I know of no way to scientifically test the influence of luck in our lives but, based on extensive, albeit qualitative, experience, I am willing to offer Lyle's Law of Luck: *The more you learn, the luckier you get.*

While I'm sure you can never learn enough to completely remove the unexplained randomness (aka, luck) from your life, I am confident that accumulated knowledge can skew the distribution in your favor. My wife and I have a favorite saying: "Dumb luck beats careful planning every time." One of the times we used it was last spring when we arrived in Lisbon without hotel reservations. Following a series of (now) amusing misadventures, we ended up in one of the greatest hotels in the city, inside the castle walls, high on a hill. Dumb luck, yes, but we had learned a few words of Portuguese, we had studied the layout of the city, and we knew what to do when we didn't know what to do. Maybe luck isn't really so dumb.

Like all of Lyle's Laws, this one isn't new. Louis Pasteur, the great French scientist—who did not, as one comedian claimed, invent milk—said, "Chance favors the prepared mind." I don't know the circumstances that prompted him to say that, but I like to envision someone saying to him (in French), "Gosh, Lou, you sure were lucky to find out that heat will keep beer from spoiling."

It is good to be lucky, but one should not count on it. Sales consultant Rick Page wrote a book called *Hope Is Not a Strategy.* Well, neither is luck. Engineers today are faced with a situation in which they really need a strategy, and luck isn't going to cut it. I am referring to the globalization of commerce. Including services. Including engineering. We have come to realize that the basic value of a good, whether it is an object or a service, is the least amount you have to pay to have it produced, plus the cost of transporting it to where you want it. It might be possible to build a bridge in China at a very low cost but if you need that bridge to cross the Mississippi River, the cost of getting it to Iowa will kill you. But the basic design of that bridge? Now that's another story.

Engineers are now facing what some other people have been dealing with for several years—the fact that if a task can be defined, segregated, and digitized, it can be sent anywhere in the world to be done. It can be and, increasingly and inevitably, it will be. The value of that task will be defined by what has to be paid to have it done plus the minuscule cost of transmitting the results back to where they are needed. *The World Is Flat*, a book by Thomas Friedman, provides a vivid picture of this situation—one that has overtaken accountants and is starting to affect engineers.

What are engineers to do? Well, if they are lucky, their jobs won't be affected—at least not until after they have retired. Ah, but luck is not a strategy. But ah, again. You *can* be lucky if you pay attention to Lyle's Law of Luck: *The more you learn...* We have always said that engineers have to keep on learning throughout their careers but the new "flat world" lends a special urgency to this message. Not only must engineers learn new things, they probably also need to learn new *kinds* of things. There isn't much payoff in learning how to do something if someone in another part of the world where wages are much lower is learning to do the same thing *and* that "thing" can be defined, segregated, and digitized.

So what *should* engineers learn? I think there are a few clues. If a task that can be defined, segregated, and digitized is likely to be available at a low cost, someone who can define, segregate, and digitize is going to be pretty valuable. So is the person who can de-digitize, de-segregate, and apply the tasks once they are completed. Some tasks can't be segregated, e.g., the application of judgment, based on local values. And I don't think you can outsource innovation. Or the management of innovation. I think that those engineers who learn to do these kinds of things are going to find they are pretty lucky.

While I have been writing this as advice to engineers, I don't want my non-engineer readers to think they are off the hook. Virtually every field of professional services in every developed nation of the world is experiencing the same pressure. If the service you provide is portable, you need to develop a strategy. And it shouldn't depend on luck.

In closing, I need to tell the rest of the story of our kids' object lesson in gambling. After winning the 13 nickels, I decided to stretch my luck and feed them back into the machine. As I recall, I "won" one more time before the nickels were gone. Holding my breath, I inserted one more nickel and pulled the handle. Not a jingle. Not a ching. Lesson complete.

Get it?

LYLE'S LAW OF

ACTION

Do something, even if it might be wrong.

IT WAS A WINDY DAY. No, not just windy—it was WINDY! I was learning about the famous Santa Ana that assails Southern California from time to time in the late fall and winter. For some reason—known to meteorologists, I'm sure, but not to me—the cool and generally moderate winds off the ocean are replaced for a few days each year by a hot, dry, intense wind blowing from the desert. This gale causes or compounds all sorts of mischief, not least of which is a series of treacherous and damaging brush fires that wreak destruction on both nature and the works of humankind. For the USS *Norton Sound,* my home for some three years, it was at least an inconvenience and had the potential of being much more.

Coming into port on a Santa Ana day, we found the wind blowing us away from the wharf where we were to tie up. Even with the help of two Navy tugboats, with their massive engines and propellers, we were not able to move against the enormous force of the wind. Our captain, seeing that we were not going to make it to our assigned location, moved us to another wharf where the wind was blowing onto our bow instead of at our side. The only problem was that the space at this wharf—due to the presence of other ships—was only slightly longer than the *Norton Sound.*

Still fairly new to the ship, I had just been promoted from line handler to "talker." My job was to wear a set of sound-powered phones and stay close to the officer in charge of the fantail, relaying any messages that he might receive or need to send. As we approached our new wharf, I received an urgent message from the bridge. The captain wanted to know if our stern was going to clear the ship behind us. I relayed the message, but instead of getting the expected reply, I found an uncertain ensign, unable to decide if there was room for us or not. Time passed, the officer remained speechless. The message from the bridge was repeated, this time with greater urgency. Finally, concerned that responsibility for generating a reply would somehow devolve upon me, I asked if I could just say we would be okay but it is close. He approved, I told the bridge, and we moved into place with, much to the relief of the ensign and his talker, a few feet to spare.

This experience, and others like it, inspired Lyle's Law of Action: *Do something, even if it might be wrong.* Here was a case where one person's inaction prevented our whole ship from taking action that was necessary for our well-being. But what if he (we) had been wrong? I don't know, but I do know that the *Norton Sound* could not stay where she was. The captain needed eyes in the back of his ship and he didn't have them. Better to see imperfectly than not to see at all.

Apropos the Law of Action, I have long been troubled by a dictum that is attributed to Hippocrates and is sometimes thought to be part of the Hippocratic oath, "First do no harm." Isn't it better for physicians to prescribe a treatment they think might help, even if they know there is a possibility that it might do harm? If I am dying, I want my doctor to try a treatment he thinks might save me, even if he knows that it might kill me.

It is often helpful to apply this law when you are solving a problem. I wrote once about a student who didn't take time to think first about how to work the problems on a test but instead jumped right into writing and calculating—usually the wrong thing. That is clearly a bad practice, but equally bad is not starting to work a problem because you don't know where to start. If you don't know where to start, start anyway. If you are going in the wrong direction, it will soon become clear. It's amazing how often a solution will suggest itself once you start to take a few steps.

When I was starting my academic career, I went to some technical meetings where I didn't know anyone. It seemed like everyone else had friends there and whenever there was a social event there would be groups that gathered and seemed to exclude everyone else. Of course, as the new kid, I was among the excluded so I found myself standing around by myself at the receptions. I just didn't know how to break in. Finally, I decided to do something, even if it might be wrong, and I went up and tried to insert myself into a group. It didn't always work. Sometimes I did it wrong. But sometimes I did it right, and I met some great people whom I still count among my friends.

As an aside, this experience taught me another valuable lesson. I guess I am an insider now and when I go to a meeting, I know lots of people to talk with. But I remember the times when I stood alone at the reception and now I go looking for the loners and introduce myself to them. Try it. It's not a law, but it is good advice.

Of course, there are times when the appropriate action may be to take no action at all. The decision to do nothing should, however, be a reasoned choice and not just the result of fear of doing the wrong thing. Many times we avoid visiting a grieving friend because we "don't know what to say." Better to say something, even if it might not be perfect.

In the end, this law is the essence of engineering. Some years ago, I gave a lecture with the rather grandiose title of "The Third Culture." My goal was to make a distinction between engineers and scientists, taking issue with C. P. Snow's assertion that they are the same. There are several differences, but the most important is that engineers, whose job is to harness nature, are required to take action, while scientists, whose task is to understand nature, are not. Scientists can just keep pecking away, approaching an answer asymptotically. Nature isn't going anywhere. But engineers have no such luxury. In the design process, we have to make assumptions, linearize the nonlinear, estimate quantities that we are not able to measure, and then *do* something. Engineers rarely find the perfect solution (seldom even the best, since "best" is hard to define), but they do generally find the optimal solution. They do it by doing something— even if it might be wrong.

LYLE'S LAW OF

GLIMMERS

Mind the glimmering at the horizon.

HARVEY DUNN IS ONE OF MY FAVORITE ARTISTS. In case you are not familiar with him, I'll give you a little background. Dunn was born on a farm in eastern South Dakota in 1884 and, at the age of 17, enrolled in what was to become South Dakota State University in Brookings. He didn't stay there for long, however. One of his teachers recognized his considerable artistic talent and suggested that he study art in a more serious way, which he did, first in Chicago and later in Wilmington, Delaware.

Dunn was of the genre of N. C. Wyeth, who was a classmate and friend and even best man at Dunn's wedding. Dunn was an illustrator for several magazines and produced battlefield drawings for the U.S. Army in France during World War I. I like his style, and I like the subjects he chose later in his life—scenes of the South Dakota prairie, an area that I know and love. I also admire his spirit and vision, which, as will presently be revealed, lead us to Lyle's next law.

I am taking a bit of a risk here, because I am going to depend on my memory, an increasingly unreliable source. As I recall, Dunn wrote about his early years on the farm and on the campus and, in describing his own awakening, used a phrase that struck me as ever-so-descriptive of a spirit that moves people—if not always to greatness—at least to greater fulfillment. He said that he was more or less content where he was, but then he could not fail to notice "the glimmering at the horizon."

What an image that conjures. A cold, dark winter night on the great prairie. A light in the house and another in the barn and one or two more at other farms, far in the distance. Otherwise, darkness. Except...except at the far horizon where something is glimmering, flickering, just beyond the familiar, awakening in a young farm boy a wish, a need, to go and find out what is there. Dunn's expression and the image it inspires lead us, as promised, to Lyle's Law of Glimmers: *Mind the glimmering at the horizon.*

We don't know just what glimmerings Dunn saw, but we know they led him to Brookings and then to Chicago and then to Wilmington and then to France. Neither do we know what glimmerings he saw and chose to ignore, but I suspect there were a few, and I also suspect that he regretted ignoring them.

But enough about Dunn. How about the glimmerings that we see?

Certainly there are glimmerings that are geographic—a faint sparkle at the horizon that suggests it might be interesting to try life in another part of the country or even of the world. These glimmerings are easy to ignore, because leaving home means giving up all that is familiar and safe for the challenge of finding a new place to live, developing new friends, scouting new stores, and just generally making a new home. But the rewards are usually great as well, not only for the people who have moved, but also for their new friends, their new employers, and for the community in general.

I suspect that most people have had a glimmer of a thought that they would like to travel or even live for a while in another country. Yet only about one-third of U.S. citizens have passports. As our world continues to shrink, we need more people to mind that glimmering and spend time in other countries developing a deeper understanding of other cultures. Students, in particular, will benefit from international experience, and university faculty need to help them get it.

There are other kinds of glimmers, of course. People glimmers, for instance. Someone who is beyond the horizon that defines your circle of friends but who seems to sparkle and would probably be interesting to know. Keep meeting people.

There are job glimmers, too. While your first duty at work is to do your job well (a future law will speak to this), it is always good to keep your eye on the horizon. There will be opportunities within your own organization and in others as well. Keep an eye out for them and, at the appropriate time, dare to peek over that horizon and check on a particular glimmer. It may be just right for you, and you will never know if you don't investigate.

Of course, not all such opportunities come to fruition. I pursued a couple of presidential glimmers and even one provost glimmer, but the sparkle flickered out either by my choice or theirs. But it was fun trying.

And there are intellectual glimmers. No matter how well read and how educated and how experienced we are, our horizons are very limited compared to the breadth of human endeavor. Whoever has eyes to see, however, will find glimmerings around the entire perimeter. The problem here is not in finding a flicker but in deciding which flicker to follow. How about some history? Ancient Rome? The Panama Canal? The Great Depression? Or economics? Or political science? Or how about learning to paint? Or studying the great painters? Learn another language. Learn more about your own language. Learn how to trick your computer into doing what you want it to do instead of what the software developers think it should.

So mind the glimmering at your horizon. And go investigate some of those glimmers. Not only might the glimmer prove interesting, but the very act of going there will push your horizon to a new location, opening up new vistas. Horizons and rainbows are similar in that way. You will never know if there is a pot of gold at the end of the rainbow because as you approach it, it moves away. Horizons move, too, and because horizons limit what you can see, the more you can move them away, the wider your vision will be. And that wider horizon will undoubtedly exhibit more glimmerings to entice you.

SIMPLICITY

Eschew complexification.

E VER SINCE I GOT MY FIRST KODAK BROWNIE CAMERA, I have considered myself something of a photographer. While this conceit was never validated by any critical acclaim (although one photograph did take a blue ribbon at the Pennington County Fair), I have had a lot of fun with the hobby.

In 1958 I acquired my first 35 mm camera and, after shooting a few rolls of black-and-white film, discovered the wonderful world of color slides. As you might imagine, a photographic zealot with three children—each with an annual birthday—as well as annual Christmases and sundry other celebrations, plus summers filled with vacation and other travels, plus flowers, trees, and bugs, can produce a lot of slides in 40 years. Hundreds of slides. *Thousands* of slides. Then comes the question that must certainly have arisen right after the invention of the daguerreotype: what does one *do* with that slide collection?

Fortunately, technology has come to my rescue. With a computer, a scanner, and a CD-ROM burner, we can reduce those boxes of slides to a few billion ones and zeros and zip them on disks that can then be distributed to everyone with the slightest interest in the photographs to do with as they will. So with the exercise of some willpower, we reduced the collection to some 800 slides and sent them away to be scanned.

The problem now is that each of these slides has to have at least a modicum of information attached to it so it will have meaning for the viewer when the photographer is not there to provide an explanation. My wife, who is heavily invested in this project, volunteered to do a lot of this labeling if I would just explain to her how to use the appropriate software package. The explanation went quite well, I thought, until right at the end when I said—unwisely it seems—"Fortunately, the software is pretty intuitive." My wife replied, with perhaps just the slightest chill in her voice, "Maybe to you."

The lesson—one that I have learned a thousand and one times and forgotten a thousand—is that different people have different kinds of intuition and different tolerances for dealing with complexity. And the conclusion to be drawn from that lesson is that, to the extent possible, complexity should be diminished. I will go so far as to codify this principle in Lyle's Law of Simplicity: *Eschew complexification.*

I was introduced to this word "complexification" in an article by *Boston Globe* columnist Ellen Goodman, in which she described a "tooth-cleaning system" that comes with an instructional DVD. Good grief. A complexified toothbrush. Mathematicians will point out that complexification has a more benign meaning as well—the mapping of a set of real variables into a set of complex variables— but here I want to deal with the nonmathematical meaning, i.e., making things more complex.

First, let us consider what this law tells a design engineer. Engineers have for many years acknowledged, if not always applied, the KISS principle—Keep It Simple, Stupid. Generally, this is taken to mean that we should use the fewest components, the simplest algorithms, the fewest lines of code, etc., often to improve reliability and/or reduce cost. While these are worthy goals, they are not the main point of the Law of Simplicity. In general, the KISS principle is focused on the product—how to make the least complicated gizmo. The Law of Simplicity focuses instead on the user.

KISS says, "Don't complicate it." This law says, "Don't make it complicated." What, you ask (and well you might) is the difference? To answer this question, I turn to American Sign Language, an idiom that often seems more expressive than spoken words.

To indicate the verb "complicate," the signer places one hand above the other, palms horizontal and facing each other but not touching, and moves them in circles, suggesting that things are mixed together. To say "complicated" or "complex," the arms are placed vertically, index fingers are extended, and, as the hands are passed across in front of the face, the index fingers are bent. The sign is intended to suggest a crossing of the eyes when someone is in a state of confusion. It doesn't seem so bad to complicate things as long as they don't end up complicated or complex. In designing a device or system, it might be better to complicate it if, by doing so, you can make it less complex. Go ahead and mix a lot of components and algorithms together (compromise on KISS?) but hide these complications from the user, who sees only a device that is less complex.

One cause of complexification is the natural desire of marketers to add features to their product in order to differentiate it from the competition. The problem comes when the customer may not want those capabilities but has to a) pay for them and b) learn to either use them or disable them. I suspect that b) is the worse. My guess is that most users would rather have to enable a feature they want, rather than disable one they don't want. (Software designers take heed.)

When designing a product for human use, empathy is a great design tool. The summer before my senior year of college, I worked for the Collins Radio Company. My task was to design a production test panel to be used in the final checkout of one of the boxes of an avionic system. As my boss was helping me define the product, he gave me an excellent bit of advice. "Remember," he said, "that you will be spending three months learning all about what this box is required to do. The person using your piece of test equipment will spend about an hour learning to operate it." He could have added, "Eschew complexification."

Like most laws, the Law of Simplicity speaks to our lives outside engineering, too. Certainly those of us who are retired or are nearing retirement recognize that we have complexified our lives with possessions, commitments, and various entanglements and that it is time for some *de*complexification. But some complexity is good. Relationships with other people make your life more complex, but they enrich it in far greater proportion. Don't eschew your friends.

What you do want to eschew are those burdens that add complexity without enrichment. Don't lie. It's too complex to have to remember what you said. Don't gossip. It's too complex to feel that you have hurt a friend. Don't cheat. It's too complex to live with the fact that you've done something that a real engineer can never do.

We will never eliminate all complexity—nor would we want to—but we can keep it in check, both in our designs and in our lives.

18

WHALEBOATS

Steer, row, or stay ashore.

WHEN MY THIRD-OLDEST BROTHER graduated from high school in 1943, he immediately enlisted in the U.S. Navy, following my oldest brother, who joined the Navy in 1941, and my second brother, who joined the Army Air Corps in 1942. He went to boot camp in what always seemed to me to be about the least likely place for a naval station, Farragut, Idaho. While Farragut is far from the sea, the naval training station was on the shore of Lake Pend Oreille, so I presume the "'boots" were required to engage in what was at that time a Navy rite of passage—rowing a whaleboat.

Clifford died a few years ago, so I will never know if he actually had to get in to a whaleboat. I will always remember, though, a book that he either brought or sent home that implied he did. The book was titled *Sailors in Boots* and was a collection of drawings that chronicled the vicissitudes of naval boot camp, something that I would personally experience some ten years later. One drawing showed about twenty recruits hopelessly tangling their oars as they attempted to get a whaleboat away from the pier. It was a funny but not encouraging sight.

I have thought of this picture many times over the years and have come to think of a whaleboat as a metaphor for an organization. In this case, one that has problems getting organized and seems just to make a mess of everything. But a whaleboat can also be a model of organization and efficiency. There is a coxswain and there are some rowers but, significantly, no passengers. Everyone is doing something, be it right or wrong. Let's use this as the visual expression of Lyle's Law of Whaleboats: *Steer, row, or stay ashore.*

Now, this seems a lot like the adage, "Lead, follow, or get out of the way," but with a nautical twist. There is a critical difference, however. The act of following is, or at least can be, a passive activity. In contrast, there is no such thing as a passive rower. Rowing is hard work, requiring a considerable amount of effort if the craft is going to make any progress. Furthermore, all the rowers must be synchronized. If one decides to row a little slower or a little faster or would prefer to set a different course, chaos will reign.

In our jobs, we have probably all known people who seemed to think of themselves as passengers on the company whaleboat. They might be willing to follow, but they have little interest in rowing. They show up every day and do enough to get by, but never really bend their backs to the task. What I find fascinating is that they often don't even realize they are not contributing very much to the success of the company. Indeed, they don't even seem to understand the connection between the company's success and their own (the subject of a future law). If they did, I think they would either get to work or get out.

Advice on this subject was provided by Elbert Hubbard a hundred years ago. He wrote, "If you work for a man, for heaven's sake, work for him...and stand by the institution he represents," or go find another place to work. (I should note that Hubbard wrote this at the beginning of the twentieth century and, even though married to a noted suffragist, followed the contemporary standard of using the masculine gender in describing a boss. Having worked for women in the last few years before I retired, I would have written it differently.)

There are also people who are very hard workers but who work only in their own interest and not in that of their organization. They are probably less prevalent in industry than in academe where faculty members become de facto free agents after they have achieved tenure. It takes only one, though, to slow the organization's progress seriously and have it limping along like a whaleboat with all of its rowers pulling hard, but with one who is out of sync with the others.

In addition to the rowers, a whaleboat also has a coxswain. It needs one. It can't have two or three or any number other than one. The coxswain handles the tiller, determining the direction the boat will go, and also calls the strokes, coordinating the efforts of the rowers.

I don't think I have ever seen an organization that officially had two "coxswains," but I have seen at least one in which one member tried to run some things independently of the boss. It didn't work. The organization was trying to go in two different directions and, as a result, went nowhere.

We would all be wise to stop from time to time and consider the various boats in which we are traveling. One of these, of course, is our work. How does it feel? Are you willing to put your back into it? Or do you feel like a passenger? If you are pulling your weight, are you synchronized with the others in your organization and all pulling in the same direction? If you aren't satisfied with the answers to these questions, you probably need to give some serious thought to how to get tuned up—or change boats.

And maybe you're the coxswain. How is your crew doing? Are you able to provide the leadership that is needed? Remember, this is the twenty-first century and you aren't a galley master with a drum and a whip and a license to throw overboard anyone who isn't doing well. You need a more delicate hand on the tiller and a friendly, albeit firm, voice in calling the cadence. You also need an open ear because sometimes a crew member will spot a hazard or an opportunity that you have missed. And you need an open mind, being willing to change course if necessary, even if the course change wasn't your idea. In the end, however, you are the one responsible for the direction and distance traveled.

Outside of work, most of us are involved at varying levels in other organizations such as professional societies, churches, civic organizations, or social clubs. I even belong to an investment club and the "Amigos de Español," a group that gets together to speak Spanish. Not all members of such organizations need to be heavily involved, but everyone needs to be at least lightly involved. Everyone needs to pull on an oar, even if some don't dip it very deeply into the water.

So check your boats. Some you steer. Some you row. All are taking you somewhere.

LYLE'S LAW OF

CERTITUDE

The more certain you are that you are correct, the more imperative it is to consider that you might be wrong.

W E WERE HAVING A DISCUSSION about a strategic decision that would have a significant effect on our organization and on some of our personnel. One of the participants held some very strong views, but the decision had pretty much been made and it was not in line with what he wanted. Nonetheless, he was quite adamant—and vocally so—that his way was the right way. At length, someone asked him, "Have you ever considered the possibility that you might be wrong?" His answer: "No, absolutely not." From then on, I essentially discounted his opinion and put more stock in the position of the people who had, indeed, considered that they might be wrong, looked for more evidence, reconsidered the question, and still returned to their original decision.

From this experience, and others like it, arose Lyle's Law of Certitude: *The more certain you are that you are correct, the more imperative it is to consider that you might be wrong.* This is tough advice to take, because it goes completely against human nature. When I really have confidence in my conclusion, shouldn't I stop looking for countervailing evidence and devote my energy to defending my position? Certainly every instinct would drive me in that direction. But certainty can be blinding and proceeding blindly is not good whether you are driving an automobile or devising a strategy for a company.

As I consider the application of this law to engineering, I find that it seems to be pretty well observed in the mainstream of our profession. Engineers are in general a cautious lot and most decisions are weighed carefully and repeatedly. At the level of what might be called "ancillary" considerations, however, there might be a need. Have you ever heard anyone say something like, "I'm 100 percent confident that this…," for instance, "…won't pose an environmental problem"? Or "…is completely safe"? Or "…is unsinkable"? When someone is that certain they are correct, it is *really* imperative that they consider that they might be wrong. An iceberg? Who knew?

In our private lives, a good place to start applying the Law of Certitude is to our prejudices. We all have them. The only way we can support a prejudice is to be *absolutely certain* that all, *all*, of "those people"—be they a racial or ethnic group, the faithful of a particular religion, members of the yacht club, or supporters of a particular political party—are stupid or dishonest or venal or dirty or lazy or greedy or what have you. Applying the Law of Certitude and exploring the possibility that you might be wrong about that will inevitably show that you are. The resultant counterexamples will destroy the generalization upon which the prejudice was based. Sorry about that.

While prejudices are generally considered to be bad, moral positions are usually thought to be good. Unless, of course, your moral position conflicts with my moral position. The issues that are so polarizing in American society today—abortion, gay rights, the death penalty, gambling, etc.—are generally defined by strongly held moral positions. We would probably all benefit, individually and as a society, if everyone who is really, really sure they are correct would consider the possibility that they might be wrong. Even if minds were not changed, the exercise would undoubtedly produce some empathy and understanding that does not currently exist.

Perhaps the most important message to be garnered from the Law of Certitude is a counsel toward humility. Certitude carried to extreme is hubris, a condition characterized by arrogance and self-adulation. Not very helpful—or admirable—characteristics.

My etymological peregrinations in pursuit of a more thorough understanding of the word "hubris" resulted in an interesting circular set of definitions. Under the entry on hubris, Wikipedia suggests that the reader also see "victory disease," which I had not heard of before. This is essentially military hubris, where an army, having won one or a few great battles, becomes overconfident and arrogant. Examples cited are Napoleon's troops at Moscow and the Japanese military in the early months of World War II. This arrogance and overconfidence leads to complacency, which ultimately leads to defeat. In turn, it strikes me that complacency is one source of the certitude that keeps us from examining our strongly held conclusions. What, me worry?

Thus, certitude leads to hubris, which leads to overconfidence, which leads to arrogance, which leads to complacency, which leads to certitude. It is a circle that needs to be broken.

This law, like any, can be misused. It should not be invoked to justify waffling or inaction. Engineers cannot afford the luxury—if that's what it is—of continuing to say, "But on the other hand,..." because our goal is to do the right thing at the right time and that means we can't delay a decision indefinitely.

I also considered saying that the law should not be used to continually question decisions that are already made and in the process of implementation. It seems that once the train has left the station, the wise course is to pour on the coal and head on down the track. I concluded, however, that this might not be so smart. The more certain I am that the track is clear ahead, the more imperative it is that I check to see if any bridges are out. I may have to stop the train, but that's better than the alternative.

As I write this column [Spring, 2007], the events in Iraq continue to unfold, page by bloody page. The reader could be forgiven for thinking that this law was written as a result of this situation, but such is not the case. The Law of Certitude has been on my list of future laws for quite some time and it is being written now because it just happened to float to the top. At this time, I don't think it would be helpful to use the law to excoriate anyone for decisions made in the past. I do, however, hope our leaders will be assiduous in applying it as they determine our course for the future.

20

GRANDCHILDREN

Design every product as if your grandchildren were going to use it.

I T HAS BEEN SAID THAT DESIGN, in its fullest sense, is the essence of engineering. This fascinating process of defining a product, apply-ing the tools of analysis and synthesis to create what has never been, evaluating an array of possible implementations of this idea, and finally choosing and implementing the "best" of these is little understood by non-engineers. It is, however, our very life.

Over the past three or four decades, engineering educators have gradually, if not always cheerfully, embraced the teaching of design in the undergraduate curriculum. "Encouraged" by ABET, the organiza-tion that accredits engineering degree programs, educators started in-troducing open-ended problems with multiple solutions and eventually teaching—or at least modeling—the process of evaluation and selec-tion. It is this final step—evaluating various designs and deciding which one to implement—that is the subject of Lyle's Law of Grandchildren: *Design every product as if your grandchildren were going to use it.*

A three-word description of the design-and-produce process is, "Innovate. Evaluate. Actuate." To innovate is to come up with new combinations of whatever it is you work with: wheels, amplifiers, gears, sensors, chemicals, pumps, airfoils, fan blades, whatever. To evaluate is to determine the parameters by which the goodness of this new combination will be judged and to assign value to each of those parameters. To actuate is to compare and balance those values and make a choice as to what the final design will be and then to build and perhaps produce this new product. It is at the intersection of the last two steps, evaluate and actuate, that Lyle's Law of Grandchildren finds its application.

What parameters will you use to evaluate your design? I would wager that one parameter will be cost. Then there are speed, power, energy efficiency, safety, environmental impact, etc., etc. Talk about apples and oranges. And grapefruits and turnips and kumquats. How do you trade off one of those against another?

Well, the value judgment often comes down to a values judgment. And that's where Lyle's Law of Grandchildren comes in. Make these decisions as if your grandchildren will be using the product. Of course it doesn't have to be grandchildren. It could be your parents or your cousin George or your Aunt Ida. The important thing is that you consider the user to be a real person who really matters to you and is not just an abstraction. Because the product will be used by—or at least will have an effect upon—lots of real people.

The first reaction to this principle is probably, "Well, if this is for my grandchildren, I'll concentrate on making it completely safe." Unfortunately, that is likely to make it so expensive that your grandchildren can't afford to use it. Or so cumbersome that they won't *want* to. So of course we are back to making trade-offs. Trade-offs that are informed, however, by the assumption that the product will be used by someone near and dear to you.

We hear a lot about corporate greed and, from time to time, about engineering decisions being made purely on economic or market-based considerations. Certainly there have been many such cases. In the main, however, I believe that most of us want to do what is right. But in the world of technology, those corporate managers who happen to have had a nontechnical education may be ill-prepared to decide what "right" is in the area of product design. They will, of course, know it is right for the company to show a profit, but that alone does not ensure long-term company success. Engineers need to be judicious in making the decisions that are within their own power but, beyond that, they also have a responsibility to provide management with a "grandchild-sensitive" analysis and to help them think in terms of real people. Is that the application by proxy of the Law of Grandchildren? Whatever, I think the general welfare is advanced if decisions are made while considering real people instead of some abstract "user."

This brings me to a corollary of this law that I call the Grandma Rule. Everyone is familiar with the Golden Rule, variously stated in various cultures, but usually seen as "Do unto others as you would have them do unto you." Pretty good advice. The Grandma Rule, however, holds you to an even higher standard.

Do unto others as you would have them do unto your grandmother.

Most people, I think, would be willing to suffer a level of indignity or inconvenience that they would not want their grandmother to experience. This tightens up the specifications on "Do unto others…"

We are living in a time when civility seems to be on the decline. While it may be argued that it has always been thus, even those who so argue will, I think, agree that interpersonal relations could use some improvement. Road rage incidents are quite common and lead to a number of traffic deaths every year. The "shock jock" radio programs and the television shouting matches that are supposed to pass as dialogue are models of incivility. (I would, however, suggest that they have had a beneficial effect in that they have driven many people away from the electronic media into the arms of a good book.) Yet they continue to prosper and seem to attract audiences, probably of people who will emulate the behavior they hear and see. Surely this is not the way the performers or their audience would like their grandmother to be treated. Can we get them to apply the Grandma Rule?

In the end, of course, the Law of Grandchildren and the Grandma Rule do not provide objective procedures to be followed. A design involves dozens of small decisions made over a period of time. Our interpersonal relations are dozens of small actions taken day after day. The important thing is that these decisions and actions be taken as if they had a direct effect on some real people who are dear to us. Because eventually they will.

MUTUALITY

A group can only succeed if its individual members succeed. And vice versa.

I N THE SUMMER OF **2008,** somewhat to my surprise, Dorothy and I celebrated our golden wedding anniversary. It was a surprise not because we had been married that long, but simply because it had *been* that long. Fifty years—and a wonderful fifty years it has been—is a long time. Half of a century. One-twentieth of a millennium. Does this give me a license to expound on what makes a successful marriage? I don't think so. I'll leave that to the psychologists and sociologists. It does, however, give me an opportunity to talk about one characteristic of a marriage that seems to apply to any kind of partnership. The result will be Lyle's Law of Mutuality, which shall be revealed shortly.

I was first introduced to this notion back when I was in the Navy and, having completed boot camp and nine months of technical school, reported aboard the USS *Norton Sound.* Like most other members of the crew, I was assigned to a position in what is known as the "Sea and Anchor Detail," which defines everyone's job when the ship is entering or leaving port or an anchorage. Since I had yet to achieve the status of petty officer, my position was as a member of a line-handling party on the ship's fantail. "Party," as used here, is a curious naval term that means group or team but did not in any way describe our activity. Line handling was no party. It was hard work and, occasionally, quite dangerous. At such times, the boatswain's mate would say, "Okay, boys. One hand for the ship. One hand for yourself."

What did he mean? Well, it took me a while to grasp the full significance of this advice, but I finally deduced that he was telling us to take care of ourselves while also working for the team. A sailor who dedicates himself totally to the ship without any regard for his own safety won't last long in that environment. Some accident will befall him—a parted line, a leg caught in a coil, any number of things. And then, not only is the sailor in pain or worse, the ship has lost a sailor. The sailor is hurt. The ship is hurt.

This principle applies to any collection of people, be it two or ten thousand. Lyle's Law of Mutuality summarizes it this way: *A group can only succeed if its individual members succeed. And vice versa.* Let me comment briefly on a group of two—a marriage.

One might conceive of a marriage in which one of the parties totally suppresses their identity and dedicates all of their energy to the partnership, but I wouldn't expect it to be a very successful or a very interesting marriage. I think the best marriage is a partnership of two individuals—each competent and self-reliant in their own right but dedicated as well to their joint mutual success.

But I said I wasn't going to expound on what makes a successful marriage. Let me turn instead to our work. If you are the boss, what kind of employees do you want? There may have been a time when the boss might have asked for employees who put the interest of the company always and far ahead of their own. Can that work? I was tempted to say that it might if the work is simple manual labor, employing workers who are interchangeable and replaceable. But even here, the workers have to take care of themselves with food and water and occasional rest or they—and their employer—will have a problem. Productivity will decline until the worker has to be replaced—a not inexpensive process in itself.

While the Law of Mutuality holds for manual laborers, it applies even more strongly to professional workers such as engineers. One hand for the ship, certainly. For the professional, this means more than "a full day's work for a fair day's pay." It means accepting and working toward achieving the goals of the organization. It means exercising the duty of care, protecting the intellectual property, trade secrets, and know-how of the company. It means having a loyalty that admits honest and constructive criticism but not mean-spirited bad-mouthing.

And one hand for yourself. Outside of work, live a life. Enjoy your friends and family. Have a hobby. Go to a party and forget about work for a while. And continue your education. Many companies used to—and I suppose some still do—support only those education programs that were directly related to the employees current job. A sort of unenlightened self-interest position. The attitude toward education is different today, with the more progressive companies realizing that virtually any education is better than no education at all and, if the employee will learn, the company will provide support.

You also need to have one hand for yourself while you are at work. A few sentences ago I said that you need to work toward achieving the goals of the company. Well, you also need goals of your own and you need to work toward reaching them. Of course, while your goals will not be the same as those of your employer, neither should they be contrary to them. If they are, you should probably be updating your résumé.

At the same time, managers have to respect and, indeed, encourage their employees to work in their own interests as well as in the interest of their employer. Not always easy, but, in my opinion, essential. In my own experience, I watched—and I hope helped—associate deans mature and become more capable until they went off to greater responsibilities and rewards. I missed them when they left but I'm sure they had contributed more to the school than if they had not been growing as they worked.

In the end, as is usually the case, it is a matter of balance. A group, be it a company, a department or a line-handling party, is a collection of individuals working together in a situation where the goals of the individuals must be balanced with those of the group. If the balance is upset in either direction—if the sailor pulls on the line with both hands but fails to hold on to the mast, or if he clings to the mast with both hands and doesn't help with the line—the success of the group will be severely diminished, if not lost altogether. Mutuality—simultaneously working toward one's own goals and toward the shared goals of the group—will help assure the attainment of them all.

CONVENTIONALITY

Question the conventional wisdom. What "everyone knows" might be wrong.

ENGINEERS ARE PROUD OF THEIR PATENTS. A patent in your name is a testimony that the technical community has examined your idea and has certified—with a red ribbon, even—that this idea is original and is deserving of the protection of the government for a period of years. A patent is an intriguing document, couched in rather arcane legal phrases that are designed to be as unambiguous as possible so that the claims made are properly restrictive but do not infringe on other claims made in other patents. I wouldn't recommend patents for recreational reading, but every engineer should run through a few just to get the feel for how they look.

In the early days of my career, I was granted two patents: one for a remote control system and one for a ferroelectric memory element. On the latter patent hangs an interesting story.

A ferroelectric may be considered the electric analog of a ferromagnet. While a ferromagnetic material can be magnetized to become a permanent magnet, a ferroelectric material can be electrified and become a permanent electret. (I'm not making this up.) In the 1960s, computer memories were based on little ferromagnetic toroids (doughnuts) that could be magnetized in one direction or the other to store a one or a zero. It was not too great a stretch to design a dual structure using ferroelectrics and configure it to represent a one when polarized in one direction and a zero in the other. So I did. The work was reviewed on campus and a patent was applied for and, eventually, granted.

Now the other thing that engineers—especially assistant-professor-type engineers—like is publications, so I wrote up this marvelous invention and submitted the paper to a well-known journal. It was rejected. The reason for rejection was a statement by a reviewer that, "it is well-known that fringing fields can not cause switching in ferroelectric materials," therefore, according to said reviewer, this thing could not work as I said it did. Well, I was disappointed, of course, but I was being pulled in other directions at that time and I was advised not to get in a writing contest with the respected physicists of this respected publication so I dropped it and went on to other things.

Fast forward about twenty-five years. In the mid-nineties I ran across a paper reporting on a device that exhibited exactly the behavior I had observed in the late sixties. Apparently fringing fields *can* cause switching, in spite of what used to be "well-known." There's a law in that story—Lyle's Law of Conventionality: *Question the conventional wisdom. What "everyone knows" might be wrong.*

I state this law not because I did follow it but because I probably would *not* have. Had I, in my literature search, learned that "everyone knows that fringing fields can not cause switching in ferroelectric materials," I doubt that I would have tried building this device. While the course of human events was little changed by my invention (although the patent is still cited from time to time), it would have been a pity if I had not had the pleasure of building and testing it and, not incidentally, receiving a patent.

So when and where should engineers use this law? Certainly in the early stages of product definition. Fifty-odd years ago, everyone knew that computers would only be used by people who were technically literate or at least involved with numbers: engineers, bankers, statisticians, actuaries, and the like. Housewives? Ha! Artists? Double ha! My friend Flo? Get real. But a few people, like Steve Jobs, questioned the conventional wisdom and saw the world in a different way. They defined a product that was complex on the inside but intuitive and relatively easy to use on the outside. Today, of course, personal computers are ubiquitous throughout the developed world and it is a major goal of developing countries to obtain computers and teach their people to use them. What everyone "knew" was definitely wrong.

There is a similar situation with the limits of technology. I haven't heard much about this recently, but in the last half of the twentieth century there would occasionally appear an article—or even a whole special issue of a journal—devoted to determining just how far technology would be able to take us. In that case, it wasn't "everyone" who knew, it was "the experts," and the same principle applies because everyone tends to agree with the experts. But of course those limits kept being broken and some expert would have to write a new article defining some new limits. I presume there are limits to technology but I also suspect we are limited in our ability to know what those limits are. So don't let your imagination be restricted by what everyone knows. Remember, there was a time when it was believed that a heavier-than-air craft could never fly under its own power—and certainly not when it was powered by its passenger—and, later, that an airplane could never fly faster than the speed of sound.

There are non-engineering applications of the Law of Conventionality as well. For instance, stereotypes are nothing more than what everyone knows about a particular group of people. Women can't be good engineers. Right? Fortunately, we have questioned that bit of conventional wisdom and found it wanting. No stereotype can stand up to a critical analysis.

This law is also useful in dealing with ex cathedra statements like, "That would cost too much," or "We have to build that out of titanium." While the speaker isn't explicitly saying "Everyone knows…" the statement is made with such authority that the listener accepts it as the conventional wisdom. Question it. Diplomatically, of course. Such speakers do, indeed, often occupy "the chair" or have considerable experience. You are, however, entitled to know the reasoning behind such a statement and will serve everyone well by asking for it.

On the other hand, we have to recognize that the mythical "everyone" is often correct and there is a lot of good conventional wisdom out there. Observing this law doesn't mean questioning everything to the point where nothing gets done. Just be alert when you hear, explicitly or implicitly, "but everyone knows…"

Write it down.

MY COUSIN MARGARET—known affectionately as Maggie—was born in 1916. Maggie was educated in a one-room country school, probably to about the eighth grade, although I'm not really sure. I know she didn't go to high school. Somewhere in her education, though, she learned to write with a clarity and concision that made her letters a joy to receive and read. Her vocabulary was limited and her grammar could be a bit creative but she kept us up to date on the life that she and Orval shared in Iowa as well as the various trips they took around the United States.

Maggie was also an inveterate diarist. I'm not sure when she started, but she had a collection of spiral-bound notebooks in which she had recorded anything that she found of interest. If someone would ask, "When was nephew Bill born?" she would go find her diary and come back with, "January 23, 1937." Pause. "At 7:33 in the morning." Pause. "In Marshalltown hospital and he weighed seven pounds and nine ounces."

Maggie died in 1998 and Orval in 2001. Unfortunately, no one knows what happened to the diaries. I suspect that as she neared the end of her life, she disposed of them herself. It could well be they held a level of personal detail that she didn't care to have other people read. However, while she didn't leave her diaries, she did leave us a model that we should heed—a model that is embodied in Lyle's Law of Records: *Write it down.*

I have never been clear on just when one should write something down versus when they should write it up. I suppose it is related to the burning down or the burning up of a building. Down or up, the result is the same. In the burning case, the building is gone.

In the writing case, there exists a compensation for the limitations of our memory.

Some people have better memories than others. A friend of mine who was noted for being somewhat "veracity challenged" was said to have such a good memory that he even remembered things that never happened. Creative memory notwithstanding, we all have imperfect memories, either forgetting information or remembering details that are not quite an accurate description of past events. There is only one solution: write it down.

For engineers, this law must be inviolable. There are several reasons. Clearly, the most important is to help ensure the quality of your designs. Nothing is of greater importance to engineers than the quality of their work, affecting, as it does, the health, safety, and welfare of the people who use the resultant products. The results of a test or measurement, the connector pin number where a signal was sampled, a suggestion made by a colleague...Of course you will remember them. Or not. Write them down. If you don't, you may forget them. Or perhaps even worse, you may remember them incorrectly.

Another reason for writing things down is to provide a legal trail. The work of engineers often results in patents. If a patent is contested, one very important datum is the date on which the idea was first conceived. The engineer's logbook should record that information in an uncontestable fashion. When I reported for my first engineering job with what was then the Collins Radio Company, I was issued a logbook and instructed how to use it. One bit of advice—actually more of a command—was that if I recorded anything that could remotely be considered to be patentable, not only should

I sign the entry, I should have a colleague witness it as well. And, of course, every page was numbered, dated, in ink, no erasures, blank areas crossed out, etc. The idea is to produce a record that is as unimpeachable as possible.

Writing provides another benefit beyond augmenting memory or providing a legal record. For me, anyway, it is a significant aid to the reasoning process. With a nod to Samuel Johnson, writing, like the prospect of hanging, wonderfully focuses the mind. Since it takes more time to record a thought than it does to think it, there is time for evaluation. Since it takes effort to write it down, we may well decide that some thoughts aren't worth recording. (Be careful, however, because it's hard to know beforehand what will be important.) And when we record our thought process, we are assisted in developing a logical progression of inferences and provided a means of testing them against observed—and recorded—facts and earlier conclusions. I have had the trying experience of reaching a conclusion through the usual thought process and then, while writing it down, finding that it just didn't hold up. I have also had the pleasant experience of solving a seemingly intractable problem by writing about it. I prefer the latter.

Today, of course, we have the computer. It provides a means of recording our thoughts and our activities that is much faster and far more comprehensive than a paper logbook. I conducted a limited, unscientific sampling of current practice in engineering companies of different sizes and learned that some adhere to the practice of using formal written logbooks while others use computer files and email for record keeping. In every case, however, the Law of Records was being obeyed.

Personal journals are also a good idea—if you can develop the discipline. I have not done well at keeping a record while I am at home but I have kept a travel journal for many years. There probably isn't much practical value, but it is a pleasure to go back and read about what I was doing and where I was doing it on a particular day twenty-five years ago. As I get older—as we all must if we can just manage to avoid the alternative—the imperative to keep a personal journal increases. There is a corollary to the Law of Records. As yet unproven but experimentally demonstrated, Lyle's Law of the Conservation of Memory, *Every time we forget something that happened, we remember something that didn't*, applies increasingly with increasing age and can be mitigated by good records.

Writing is hard work. Like most hard work, however, it is usually well rewarded. Whether it is an informal but precise engineering log—either paper or electronic—or a formal paper for publication, or a personal journal, writing is worth the effort. The result may be archival or it may be as transient as Maggie's diaries, but it is useful. Write it down. Or up. Whatever.

GENTLENESS

It is safer to err on the gentler side.

DEANS HAVE MANY DUTIES and most of them are quite pleasant. There are degrees and awards to be handed out, new students and their parents to be greeted, promotions to be acknowledged, and anniversaries and other occasions to be celebrated. The dean is also the public presence of the school or college and is thus involved in all kinds of social activities and operations at the interfaces between the academy and industry or government. There are challenges, to be sure, but in general these duties are both pleasant and rewarding.

While I don't want to sound like an economist, I must now say, "On the other hand…" Yes, on the other hand, there are a few duties that are, shall we say, less than pleasant. Occasionally—rarely, one hopes—the dean must recommend against tenure or reappointment. Not a happy task. Deans never have enough money, so they have to deny some very legitimate requests to fund activities that would be very desirable. And, of course, there are the cases of student discipline and disgruntled students and parents.

The fortunate dean has one or more good associate deans who stand between the dean and the disgruntled. I was certainly blessed in this regard. Eventually, however, some cases escalate to the point where the associate dean—with a sigh of relief, I am sure—has no choice but to deliver the case to the doorstep of the dean.

One such case arrived in my office two or three days before commencement. A senior who had a very good academic record had arranged to do an independent study during his last semester and had failed to meet the expectations of his professor. The associate dean had negotiated the grade from failing to incomplete but this still left the student unable to participate in graduation. Disgruntlement abounded. The parents—alternating between tears and fury—were in my office. I reasoned, cajoled, attempted to mollify. No luck. So I decided to try taking off the kid gloves and get tough. I don't remember my exact words but, clearly, they were not well-chosen. The father offered to knock my block off. Sensing that this would not be a desirable outcome for either of us, I returned to the mollify strategy. Well, the student did not walk across the stage but did graduate eventually. The father and I, who did not know each other before, ended up having an amicable relationship. And I learned—actually relearned—a valuable lesson, summed up in Lyle's Law of Gentleness: *It is safer to err on the gentler side.*

The reader may recognize this as a synthesis of two legal maxims: *Tutius erratur ex parte mittioro,* "It is safer to err on the side of mercy," and *In dubio pars melior est sequenda,* "In doubt, the gentler course is to be followed." I prefer gentleness to mercy. Mercy is something that is granted to someone who may or may not "deserve" it and there is an implied difference in status of the grantor and the grantee. Gentleness, on the other hand, suggests a trait that is well-advised no matter what the relative status of the protagonists might be.

There is considerable room for gentleness in our practice of engineering. Few will argue, for instance, that the products of our profession have always been gentle with the environment. As we begin to understand the effects that we have engendered and to recognize that the Earth is deteriorating at a rate somewhat faster than we had thought, we need to make sure that our practices change as well. There is lingering—though weakening—debate about the cause and progress of global warming and its potential impact on weather, crops, and the level of the sea. (I am particularly interested in the level of the sea because our house is located about six feet above mean high water.) To my knowledge, however, there is no debate about the increase of greenhouse gases and not much about their origins. We have helped to engineer our way into this mess and, if we strive to err on the gentler side with Earth, we can engineer our way at least partway out.

Engineers also need to be gentle with people. The change in safety standards over the past twenty years has been remarkable and virtually all of our products are far safer than we would have thought possible. Most of this has been due to external regulation, and engineers run the risk of leaning on regulations to make sure our products are safe. But our design decisions can't be solely determined by what we are *required* to do. As engineers, we know that we will not hit the design target dead on, so when we err—as we will—we need to err on the gentler side. The side of safety.

Of course, it is in our interpersonal relationships where the Law of Gentleness finds its most ready application. We interact in both word and deed with family, friends, colleagues, and a host of people with whom we come in contact for a brief time and may never see again. Ironically, we are often least considerate of the people with whom we are most intimately connected, although tales of road rage provide dramatic counterexamples. With everyone, we are continually making decisions of what to say, what to do, or what to require or allow others to do. These decisions are often made rapidly—even hastily—when a little reflection might have produced a different outcome. Better to develop the practice of giving a little more thought to the process and then choosing to err on the gentler side.

So does this make you look like a wimp? Quite the contrary. The weak have little choice over how they can behave. The strong can choose to act however they wish. Really, you *have* to be strong to be gentle. If you are being shaved with a straight razor, do you want

a barber with strong hands or weak ones? Some readers may remember the advertisements for the Hastings Piston Ring Company.
They featured a big, rough-looking guy with a lantern jaw and a five
o'clock shadow who was always engaged in some act like holding a
baby or petting a puppy. The caption was always, "Tough, but oh so
gentle." The message was clear: if the rings were not tough enough
to resist wear and breakage, they couldn't be gentle with the cylinder
walls. I don't know whatever happened to the piston ring guy but he
was a great role model.

There is an important implication in the Law of Gentleness:
the understanding that we will, indeed, make mistakes. Sorry, but
we will rarely get it exactly right, whatever "it" might be. We will
err, but, if we are smart—and tough enough—we will err on the
gentler side.

POSSIBILITY

If it can happen, it will happen.

O NE PLEASANT BUT WINDLESS SPRING DAY, my wife and I were aboard our sailboat *Gitana,* motoring up Eastern Bay, a branch of the Chesapeake. Dorothy was at the helm and I had gone below for something. As I came back on deck, I noticed a wisp of vapor coming up from the stern, apparently coming from the exhaust. I quickly checked the engine temperature gauge, saw that the engine was overheating, and shut everything down.

Now what? I was a reasonably capable shade-tree mechanic back before cars had a dozen computers, but my experience with three-cylinder diesel engines is limited (more limited at the beginning of that day than at the end). I will spare you the details of checking the thermostat, disassembling and reassembling the impeller housing, and flushing the heat exchanger. Finally, I did what I should have done in the first place and checked to see if the cooling water inlet was blocked. Why is it that the last thing you try is always the thing that corrects the problem? I poked a wire into the tube and wiggled it around a bit. Almost immediately there was a gush of water, carrying with it a little silver fish about the size of my finger. I reassembled the hose, and in another minute we were on our way.

Now, what is the probability of just the right size little fish swimming up the cooling water inlet of a moving boat? Clearly, it is small, but...Murphy's Law strikes again.

Most people—certainly most engineers—know about Murphy's Law. It has been stated in several ways and its origin has been disputed but, fundamentally, it says that if anything can go wrong, it will. It is a representation of what has been called the innate perversity of inanimate objects. For many years, however—even after coming eye to eye with that little silver fish—I have felt that Murphy's Law is too pessimistic. It is the negative side of Lyle's Law of Possibility: *If it can happen, it will happen.* This more general law allows for the possibility that both good things and bad things can happen, since "it" can be either good or bad. And, borrowing from Robert Frost, that can make all the difference.

The Law of Possibility can be broken down into two parts: Murphy's Law—If anything can go wrong, it will—and what we might call Feisel's Complement—If anything can go right, it will. Neither statement includes the words "every time," so an engineer's job is to make sure Feisel trumps Murphy as often as possible.

That is an interesting challenge for engineers. On the one hand, they must be almost ridiculously optimistic. Think about it. President Kennedy asked the engineering community if they could put a man on the moon and bring him back safely to Earth. The engineers said, "Oh sure. We can do that." That was optimistic. There was no place for pessimism in the Apollo Project. If the team (or, more accurately, teams) had not believed it could be done, it would not have been. They had to believe that the myriad complex systems would all work and that the mission could and would be accomplished. They had to believe that if it can go right, it will go right.

On the other hand, engineers must also be skeptical. They must believe that if it can go wrong, it will go wrong. Realizing that, they must reduce the probability of it doing so until it is as small as possible. This requires a very suspicious eye as well as a vivid imagination and a thorough understanding of the system and all the external and internal factors that can affect it. It also requires discipline. In design and development, the focus is on, "What will make this work?" You must also remember to ask, "What can make this fail?"

It is important to note that optimism and skepticism, as I am using them here, are not antonyms. Thus, engineers have to be skeptical optimists with an attitude of, "I know this will work but I have to defeat all of the forces that will keep it from working," or optimistic skeptics who say, "There are lots of things that will make this fail but I know I can defeat them and make it work."

Students need to remember the Law of Possibility, too. I remember a young woman who was struggling in an electrical engineering course (not mine) asking me, "Does anyone ever pass this course?" I assured her that, yes, people had been known to pass that particular course and even go on to pass all the other courses and actually graduate from our program. Greatly reassured, she went on to pass the course and graduate. We were joking, of course, but the fact remains that she needed to believe that it was possible—even likely—that she would pass. She had to be optimistic.

On the other hand, students need to be skeptics. They need to ask what might keep them from passing a course or, in other words, what they don't know but should. That is why I believe so strongly in written learning objectives, i.e., a list of what a student must be able to do to master a course. I can't learn for my students but at least I can tell them what they need to know. Then, if they know what they are expected to know and skeptical enough to question whether they know it, they are very likely to succeed. By the way, students, if for some inexplicable reason your teacher doesn't provide you with a list of objectives, write your own. It's a great way to study a subject and prepare for exams.

I hope no reader will parse the Law of Possibility and try to demonstrate that it is mathematically untenable. I'm afraid they would succeed. In the end, the message of the law is that Murphy does not rule. Yes, bad things can happen and if they *can* happen, then there is a non-zero probability that they will. So either make sure those things *can't* happen or, if that's not possible, make the probability of their occurrence as small as possible. And remember that good things can happen, too, and, with sufficient effort, the right knowledge, and a little bit of luck, they will.

26

LYLE'S LAW OF
ALTITUDE

Things look different from five thousand feet.

I N MY OPINION (humble, of course), one of the greatest computer applications on the Internet is Google Earth. This site, in case you are not familiar with it, allows the user to view satellite images of Earth from almost any altitude. When the screen comes up, you are looking at Earth from an altitude of 6,835.7 miles (more or less). This is not a particularly interesting view, but, with the flick of your thumbwheel, you can start to zoom in. You can also do a drag-and-drop and move to another part of the world and then zoom in some more and, depending on where in the world you are looking, move down so close that you can see cars and boats and even people.

My initial fascination was centered on seeing how close I could get to places with which I was familiar. Look! You can see Bill's car parked in the driveway! Then I moved on to places that I hadn't been. I could look at Mount Everest or fly across the Sahara or sail around Sydney. Magnificent. Finally, I discovered another very useful function. After I had zoomed down and flown around, I found that I could now zoom out and see something that is perhaps even more important than those up-close details. I could see how the place I had been examining close up was related to other places and geographic features. This zooming out to gain perspective is the metaphor for Lyle's Law of Altitude: *Things look different from five thousand feet.*

One thing you see from 5,000 feet—or any reasonable altitude or distance—is how things are related or interrelated. I was probably about twelve years old when I had my first airplane ride from the Tama, Iowa, airport, which was about two miles from our farm. As I looked down for the first time on the area where I had lived all my life, what amazed me was how close things were to each other and how everything fit together. Of course, I knew there was our farm and the Schraeder farm and the Betz farm and the Schuett farm and so on, but it was a revelation to see them all laid out below me and all fitting together. Somehow, it seemed there should be some vacant spaces, but of course there weren't.

So how does this apply to engineering? For one thing, I think engineers tend to live on their own "farm" and, while they know there are other farms out there, they can forget they are all connected. It used to be worse. Fifty years ago, marketing, product development, and production (now called manufacturing) didn't work together as they do today. It was pretty standard practice to design a product and "toss it over the wall" to production, who built it at whatever cost and then tossed it to marketing to price it to sell at a profit. No more. Now there is communication and cooperation throughout the product life cycle. Even so, a view from an even higher altitude can be useful.

It is often helpful to take a mental flight around your company and cruise over your department and then the other departments or even other divisions. If you get up high enough, you will probably see how they fit together and how what you do relates to other parts of the company. Engineers have sometimes been known to look with disdain upon the "overheaders" in the company. In case you are not familiar with that term, it is a pejorative that encompasses all the people who are not directly involved in making a profit: personnel (now known as human resources), facilities, finance, certainly legal, and probably some others. A friend of mine was president of a large company and also had a law degree. He was still an engineer, though, and continued to rail against the overheaders, presumably including himself. He was a wise man, however, and while he strove to minimize the overhead, he flew high enough to realize that every element of the company was necessary and that if any one was missing, the company could not function.

Engineers would also do well to take a flight to investigate how their product or process fits into the world at large. By flying at the right altitude, we will see that our activity will affect a whole range of things that are out of sight if we stay too close to our work. I well remember how astonished I was when I first saw a stream table demonstration. A stream table is a large platform covered with soil and tilted at a small angle. A straight channel is cut in the soil and then a stream of water is introduced at the top end and flows to the bottom. The demonstrator made a small nick in one side of the channel and then let nature take its course. It took a while, but over time that nick grew into a pocket, the pocket into a bend, and eventually into a meander that resulted in additional meanders down the stream. It strikes me now that an ant standing beside that nick would have seen only that the stream bank was eroding and moving. A bee flying above the table, however, would have seen the very significant overall effect. Be like the bee.

Altitude or distance can also be helpful in examining our lives and our actions. If you can back off a bit and see how the things you are doing—or are thinking about doing—fit into the overall picture, you might change your plans. There are two advantages to this. For one thing, the added perspective will help you predict what the consequences of your actions might be. You should be able to see down the road a bit and see the coming curves and bends and also the people you will see along the way and how they might be affected. The second advantage is that by backing off, you are able to stand where others are standing and, with a little effort, see your actions from their perspective. Oh, what a boon this is! Or can be, if we pay attention.

Of course, you do come back to Earth and you have to make your decisions on the ground where you live and work. Generally, however, these decisions will benefit from the perspective gained by taking a flight to 5,000 feet and seeing how things look from there.

27

LYLE'S LAW OF

REFLECTION

*Reflect upon—and learn from—
your failures and your successes.*

GROWING UP ON A MIDWESTERN FAMILY FARM in the 1940s was, in many ways, a pleasant experience. There were hundreds of acres of fields and forest, creeks and gullies, hills and valleys to explore, almost without restriction. There were animals of great variety. Some—like bulls and setting hens—were a bit scary but, for the most part, farm animals were fun to be around. What can compare to a newborn pig or calf or lamb brought into a warm house on a cold spring morning?

On the other hand (isn't there always another hand?) there were some onerous aspects. Most of these unpleasant features involved chores that needed to be done on a regular schedule, irrespective of weather or other pressing matters, such as visiting one's friend at the farm over the hill. Some of these chores were not so bad, some were bearable—but barely—and some were totally devoid of redeeming features. One such was the job of "walking the corn" or "walking the beans."

This simple process consisted of walking between two rows of corn or beans, searching for offending plants such as errant corn in a bean row or a despised cocklebur or buttonweed in a corn row and then destroying them utterly. Simple, yes, but imagine the mind-numbing boredom of walking a quarter mile in one direction, turning around, walking a quarter mile in the other direction, turning around...I think you get the picture. A terrible way to squander a day of one's fleeting youth.

On one such day when I was seven or eight, my dad and I were walking the corn when I—disgruntled, you may be sure—spotted an ear of corn that had been dropped in the previous year's harvest. I picked it up in my left hand and took a swing at it with my utter destroyer of weeds, a two-foot-long corn knife (that's farm terminology for a machete). In my disgruntlement, I aimed a bit too far to the left and, instead of neatly severing the ear of corn, I peeled a not insignificant flap of flesh from my unfortunate finger. Whereupon the corn-walking ended and we headed for the doctor's office.

For all his many faults (and who among us does not have quite a few?), my father was a wise and gentle man. When this sorry episode was finally concluded, I think he could have been excused for saying something like, "That was really stupid." But he didn't. He just asked, "Did you learn anything from that?" This question is one I have since asked many times—of both myself and others. It is the core of Lyle's Law of Reflection: *Reflect upon—and learn from—your failures and your successes.*

Following a failure or a mistake—not exactly the same thing, but often related—we are prone to engage in one of two actions: castigate ourselves or castigate someone else whom we might be able to blame. The former action might have some value if it results in an increase in humility with no reduction in self-confidence. The latter is not likely to produce much of anything useful. So you can berate yourself briefly, but self-flagellation grows old fast and you soon realize it is time to move on. Before you do, however, exercise the Law of Reflection and ask yourself if there is something to be learned from your recent unpleasant experience.

Such a reflective analysis will often move through the stages of what, how, and why. What happened? How did it happen? Why did it happen? The last question can be complicated but is the essence of the process. Here is a simple example:

What? My car stalled on the freeway.

How? It just sputtered a few times and then stopped.

Why? The gas tank was empty.

Why? I hadn't filled it.

Why? The gas gauge said it was half full.

Why? The gauge is broken.

We certainly learned something from that reflection and what we learned is actionable. Fix the gas gauge.

I won't pretend that the real-life problems you deal with are anywhere near as simple as that. Nonetheless, the basic principle applies. Think about it. Reflect on it. And dig deeper. Ask why as many times as you need to. If, in our example, we had asked the question only once, the suggested solution would have been to fill the gas tank. My dictionary says that "reflect" is derived from the Latin *reflectere* which means "to bend back." In a good reflective analysis we have to bend back and then bend back again and keep bending back until we understand all we can about what happened, how it happened, and especially, why it happened.

While it is not always easy to take time to reflect on our failures, it is probably even more difficult to do so with our successes. When things are going well, our instinct is to bask in the warm glow of success and then forge ahead to even bigger and better achievements. Even success, however, has a great deal to teach us.

A superficial reflection on a successful operation may lead you to conclude, "Oh, I just did everything right." Well, not likely. And even if you did, you need to explore just what was right about those things and how you can make sure you do them right again. And rightness is relative. As far as I know, nothing is perfect. One of the basic tenets of Continuous Quality Improvement is that things can always be better. Better to be improving on a successful program than one that is failing.

Socrates wrote, "The unexamined life is not worth living." Perhaps it would be flippant—even for Lyle—to say that the unexamined failure is not worth having. Or the unexamined success. I do not hesitate, however, to say that either failure or success is worth a great deal more if it is reflectively examined and the results of that reflection are used to guide us in the future.

INVISIBILITY

*Who you are is revealed by
what you do when you are
absolutely certain that no one
will ever know you have done it.*

THE GIRL STUDIED THE MARVELOUS ARRAY OF CANDY BARS. There were big fat ones with nougat centers and a chocolate coating, thinner ones of pure milk chocolate, round ones with lots of coconut, and some that were crunchy and studded with peanuts. The girl selected two of the most decadent and headed for the checkout counter. As she walked, she carefully slipped one into the deep pocket of her jacket. She paid for the other bar and walked calmly out of the store, congratulating herself at just doubling her money and telling herself, "No one will ever know."

The university student couldn't believe his luck. He had gone to the seat he had been assigned for the test and found himself seated diagonally behind the best student in the class. And he could see her paper perfectly. The test was a tough one but, with a few surreptitious glances at the desk ahead of him, the student was able to solve all the problems and get the answers he knew would be correct. He left the classroom with a smile on his face, confident of a good grade and telling himself, "No one will ever know."

The man and woman sat close together in the Manhattan restaurant booth. They had come to the city on business and, after the day's work was done, chose this secluded bistro for their dinner. It had been a romantic evening, with soft lights, a bottle of fine wine, and gentle background music. As they left to go to the hotel, they knew the evening was about to get even better. The couple were in their forties and were married—just not to each other. But their homes were hundreds of miles away and they were telling themselves, "No one will ever know."

Four individuals. Three different places and three different times in their lives. But they all have one thing in common: they believe they are invisible. And they probably are. Probably, no one will ever know what they have done. But who are these people *really*? We can get a clue from Lyle's Law of Invisibility: *Who you are is revealed by what you do when you are absolutely certain that no one will ever know you have done it.*

To be sure, when it comes to right and wrong, not everything is black or white and there are clearly several shades of gray. Very few people, however, would say that any of these actions would be considered the right thing to do or even that circumstances may have been such that they can be excused. But why shouldn't these people do these things? Why, indeed?

There are various reasons for people not to do the wrong thing. Probably the simplest and most direct is the Law of the Land. Since at least the time of King Hammurabi of Babylon, almost 4,000 years ago, societies have established codes of conduct that govern the interactions of their citizens. Basically, these were created to help ensure that the citizens were secure in their life and their property and to provide a mechanism for settling disputes. Such laws concentrate on wrongdoing, providing a reasonably clear description of what the state considers to be wrong and often prescribing the penalty for doing the wrong thing. The citizen's situation is pretty clear: commit a wrong (crime), get caught, suffer the consequences.

A somewhat more complex motivation comes from religious codes or commandments. It seems that most major religions have a set of commandments or principles that proscribe certain activities and prescribe others. These commandments are considered to be divinely inspired and are the divinity's way of contributing to human happiness and also, in most religions, assuring the loyalty of its adherents. Punishment can range from the trivial to the eternal. What is certain, however, is that you *will* be caught. Has there ever been a god or goddess that is not omniscient? It then becomes an issue of either being forgiven or of doing enough good things to balance out the bad things (sins).

A third motivator comes from what I call the clan commandments. These are the expectations, express or implied, of the groups to which an individual belongs. One such group is the family, both immediate and extended. The rules of family behavior are rarely written but they are usually well understood. Codes of ethics are written by members of a profession and are usually provided as guidance, leaving some room for interpretation. Of course, the keepers of the code are generally available to help with that interpretation but the final decision is usually left to the individual. More specific are the honor codes adopted by some schools, usually something very close to, "I will not lie, cheat, or steal, or tolerate those who do." Pretty clear. A violator of the clan commandments faces a range of sanctions from experiencing the disappointment or condemnation of fellow members up to expulsion from the group.

But now comes the clinker. If a person declines to do wrong for any of these reasons, what can be concluded about that person? Only that he doesn't want to be punished. Not too deep. We can't determine what kind of person he really is. We can only determine that by seeing what he does when he is absolutely certain that no one will ever know that he has done it.

But of course this is not about judging other people, an ungenerous activity at best. It is about judging yourself. You make dozens of decisions every day. What is the basis for those decisions? Do you obey the speed limit because you don't want to pay for a speeding ticket or because there are children around? On the other side of the coin, do you give to a charity because your name will be listed in the annual report or because you believe in what that charity is doing?

I hope this does not come off as a morality lecture. It is not so intended. My intention is to get each of us to consider what kind of person we really are and what kind we would like to be. It is particularly important for engineers because we often act independently and indeed find ourselves in situations where "no one will ever know." Then there is only one person to whom you must answer. In *Hamlet,* Shakespeare said it very well: "To thine own self be true and it shall follow, as the day the night, thou can'st not then be false to any man."

LYLE'S LAW OF

HIKING

Think like a backpacker.

ONE OF THE ACTIVITIES I ENJOYED IN THE PAST—and that I keep telling myself I'm going to do again—is backpacking. From a strictly rational viewpoint, there is probably not much to recommend the sport. It involves expending a significant amount of energy, putting unaccustomed stress on your legs and back, and often developing blisters upon blisters from shoes that you really intended to have broken in but didn't get around to. It often takes you into inhospitable territory where poison ivy lurks and bears wait to raid the camp of the unwary. Food is prepared by the hiker and, at least in my case, will never be featured in your newspaper's Sunday supplement. At the end of each day, you tuck your aching muscles into a damp sleeping bag for a few hours of fitful sleep on a thin, leaky air mattress. Other than that…

Well, other than that, you get to see stars that the city-bound never knew existed; you get to see a hen turkey lurch across the trail with a "broken" wing to lure you away from her chicks (poults, they're called); you get to see the forest in the early morning mist and watch as the rising sun turns dew into diamonds; you get to walk through a tunnel of blooming rhododendron that defines the trail along a mountain slope. And when you get back to civilization and have taken the anti-inflammatory and had a good shower and a good night's sleep in a bed made deliciously comfortable by memories of that air mattress, you bask in the knowledge that you have overcome a physical challenge and have experienced nature almost on its own terms.

As in any sport, there are several rules that the prudent backpacker observes: get in condition before you go; keep your pack as light as possible; have good maps and a good compass (today, I suppose, a GPS); carry enough water; plan your hike, then let someone know where you are going and when you will return; take periodic rest breaks. Each of those rules could be the basis for a good law but I'm going to wrap them all into one: Lyle's Law of Hiking: *Think like a backpacker.* A good packer will consider each of those rules. Let's do the same.

Get in condition before you go. Whether you are considering starting on a hike, a work project or a career, you need to get in condition. In backpacking, most of your preparation will be physical. In engineering, as in most lines of work, it is generally mental—a matter of education. Get in shape. If a job requires a bachelor's degree, consider getting a master's. If a project demands an understanding of discrete transforms, learn discrete transforms and the underlying mathematics. I have shared stories with a lot of backpackers over the years and I have never heard any say they overtrained for a hike. Many have said they should have been in better condition. In the same vein, I have never heard anyone say they had too much knowledge to do a job.

Keep your pack as light as possible. Backpackers say that the best way to determine if something is really necessary is to carry it for ten miles. In our work and in our lives—as in hiking—it is very tempting to try to provide for every possible contingency. We have to recognize, however, that those provisions add to our load and slow us down. Whether it is clutter on our desk, clutter in our brain, or clutter in a project plan, the superfluous can overwhelm the essential. A quest for security and comfort can overwhelm efficiency and effectiveness. Make no mistake, this is a delicate balance, but searching out and eliminating those things whose weight-to-benefit ratio is too high will be time well spent.

Have good maps and a good compass. I doubt that the Music Man was a backpacker but he understood this rule when he sang, "You gotta know the territory." Whether you are designing a product, starting an engineering school, or moving into a new community, you have to know your surroundings, your customers, your competition. What are the features of the countryside that might cause you to fail? What can you use to help you succeed? Exploring such questions will give you a good map of the area you are about to traverse. Then, as your project progresses, you need to continue to take bearings to know where you are and maybe even update your map. Continue to look at your surroundings. Just walking with your head down might not get you where you want to go.

Carry enough water. Once, returning from an overnighter in the Black Hills, I ran out of water a couple of hours from the trailhead. What could be so bad about taking a drink directly from that nice cool stream? Well, the short answer is, *E. coli*. Things were kind of loose for the next couple of days and I vowed never again to go light on the water. This is the corollary to the pack light rule. If something is essential, make sure you have plenty of it. What is essential? That depends on the context. In an engineering project, you need personnel, probably computing power, and time. In your life, you need, among other things, friends. Make sure you carry enough.

Plan your hike, then let someone know where you are going and when you will return. If there is one thing that has contributed most to the failure of hikes, projects, businesses, and marriages, it is lack of planning. A plan, of course, requires more than specifying a destination. A goal like, "Hike from the Georgia line to Front Royal" may be the start of a plan, but if the time allotted requires hiking forty miles a day through the mountains, it isn't going to work. A good plan must have intermediate goals or milestones plus an understanding of what it will take to get to each one. And it is essential that everyone involved knows the plan and agrees that it is workable.

Take periodic rest breaks. There may be hikers who can hike all day without a break, but why would they want to? Without a periodic breather, the hiker will suffer both mentally and physically and probably won't be able to enjoy the scenery. The same is true in our work and in our lives. Everyone has had the experience of working through the weekend or pulling an all-nighter, but making a practice of that is a formula for failure, not for success. Take some time to smell the roses. Or the rhododendrons. Whatever.

So think like a backpacker. The sport is characterized by physical and mental toughness, high efficiency, and careful and responsible planning, all characteristics that will serve us well. And the more I think about it, the more it sounds like fun. Maybe I'll see if I can find that old backpack.

ROCKETRY

Everybody needs a booster.

L IKE MOST YOUNG MEN OF MY GENERATION, I spent a few years in military service—in my case, the Navy. I came into the Navy just as they were preparing to deploy surface-to-air missiles in the fleet and I was fortunate to be trained as one of the first Guided Missilemen, a rating later changed to Guided Missile Technician. After basic training and some nine months in a technical school, I was sent to the USS *Norton Sound*, which was doing the final sea tests of the Terrier missile.

We rarely went far from our home port but almost every week we would put to sea for a day or two and fire missiles at remotely controlled target aircraft. My duties involved prelaunch testing and loading the missile on the launcher located on the fantail. During the launch, I was able to stand on one of the side decks and watch the flight of the missile. It left the launcher with a tremendous roar as the booster took it from zero to Mach 3 in just a minute or so. At the end of the boost phase, the missile and booster separated and a sustainer rocket ignited to maintain the missile's speed as it went about its business of intercepting the target.

The performance of the Terrier was incredible—we rarely missed. The airframe, the guidance system, the power system, the sustainer rocket: all were very sophisticated for the time—over fifty years ago—and they worked beautifully. It has occurred to me, however, that this advanced system could not have done its job had it not been for the work of the booster. This simple but very powerful rocket didn't last long but it got things up to speed and enabled everything that happened later. Further reflection suggested that this closely parallels the human condition and out of this comes Lyle's Law of Rocketry: *Everybody needs a booster.*

The purpose of a booster rocket is to overcome the inertia of the rocket being boosted and change its state of motion. It does this by converting chemical energy into mechanical energy and transferring that energy to the boostee. Some of this will be kinetic energy manifested in increased speed. Some will be potential energy associated with the greater altitude achieved. And some will be dissipated as heat, overcoming the friction that impedes the flight of the rocket. The human booster/boostee relation is much the same.

We often speak of human "energy." Since this isn't energy in the strict physical sense, I prefer to use the term "effort." But the result is the same: as one person expends some effort to boost another, the boostee is pushed to rise to greater heights—i.e., to have greater potential—and also to move toward a goal with greater speed. A good boost can also help to overcome the friction of the events and circumstances that hold us back. Inertia—not the physics kind, but the human variety—is overcome.

A boost may consist of some real assistance such as helping to solve a problem or providing a bit of cash or shoveling a sidewalk but it is often no more than an encouraging word to let someone know you think they can succeed. Or, as appropriate, a scolding and an admonition to change some counterproductive practices.

The Law of Rocketry was inspired by a visit to a campus with which I have been associated for the last 25 years or so. Part of that university's mission is to take students who may not be well prepared academically or whose motivation may be weak or whose life expectations are not very high and turn them into college graduates.

They don't always succeed, but they succeed more often than they fail—and much of that success is due to the campus boosters. And it seems that everyone there is a booster: the president, the deans, the faculty, the students, and even the alumni. Everyone is encouraging the floundering student. Sometimes it is a pat on the back. Sometimes it is a kick in a lower region. But the boost is always there. It is a community of boosters. Little Canaveral.

Perhaps the Law of Rocketry is a bit sweeping. Perhaps not everyone needs a booster. There may be self-made people who have done everything on their own and have never needed a helping hand or an encouraging word from anyone. If so, I congratulate them. For lesser mortals, however—such as yours truly—an occasional boost is essential.

As with any metaphor, of course, the analogy between the booster rocket and the booster person is not perfect. The rocket gives out a horrendous roar, much louder than can be imagined by someone who has never heard one go off, and gives a tremendous push for a short period of time. And it can do it only once. In contrast, human boosters are most effective when their work is quiet and little noticed by casual bystanders. They are also adaptive; they don't need to give a huge boost all the time but can adapt the length and strength of their push to the time and the circumstances. And they aren't single-use devices; they can keep giving boost after boost after boost.

Boosting is not the same as mentoring. A mentor helps you know what to do. A booster helps you believe you can do it. I suspect I am not alone when I say that one of my greatest boosters was my mother. Her education was received in a one-room country school during a series of "terms" that fit in between seasons of planting and harvest. It ended at what she estimated to be about the sixth-grade level. She knew nothing about higher education but it was her urging and her support that gave me the impetus to leave that little Iowa town and go to the university.

Perhaps the most important boosting comes from teachers to their students and from managers to the people who report to them. However, this process is complicated because the booster eventually has to sit in judgment over the boostee. It is difficult to say, "You can do it. You can do it," and then, later, hand out a grade of D or a poor performance review. Nonetheless, it must be done—and done in such a way that lack of success is seen as a temporary condition.

How best to boost? In the end, boosting is an attitude, not a technique. Once you understand that someone else's success does not diminish you, all you need to do is follow your best instincts. Go boost.

LYLE'S LAW OF

DECISIONS

Make up your own mind.

WHEN I (FINALLY) RECEIVED MY PH.D., I went to work for the South Dakota School of Mines and Technology as an assistant professor. Those were the days of growing research emphasis on American campuses, but at the Mines there was still a feeling of loyalty to the institution and camaraderie among the members of the faculty. One manifestation of this climate was the faculty lounge, a basement room where most of the faculty enjoyed lunch and coffee breaks. It was an atmosphere of liberal culture where we met with colleagues from other disciplines and discussed everything from antelope hunting to the state of the economy and certainly politics. One person who was always in the thick of the discussion was a history professor who also directed the chorus. He had a reputation as a wit (I told him I thought people were only half right) and could always be counted on for a clever comment. One day, in mock seriousness, he said, "I wish I could be a (here the reader may insert either major political party). Since I'm a (insert other party) I have all these questions and have to think about things and make up my own mind instead of having the truth all spelled out for me."

I have not specified which party goes in which slot because I don't want half of my readers to stop reading at this point and start writing their letters to the editor. I have known people of either stripe—or no stripe at allwho fail to do their own analysis or thinking but just accept the party line. This is what I caution against in Lyle's Law of Decisions: *Make up your own mind.*

When I left the School of Mines, I served for some years in a university whose roots are sunk deep in the loamy soil of the liberal arts. There, I heard a lot about the value of critical thinking and about the regrettable lack of it in our students and in our curricula. Critical thinking, we are told, is probably the principal educational goal of a liberal arts education. So just what is critical thinking? Well, like so many terms of this nature, it depends on whom you ask, but there are some common threads in the answers you receive. That said, I will add a few words to the many that have been written on the subject.

First, it should be noted that while the word "critical" is often taken to have negative connotations, that is not at all the essence of critical thinking. The result of critical thinking can, of course, be a negative evaluation, but it can just as often be positive. Simply put (but I hope not too simply), critical thinking is the process of making up your own mind in an orderly, objective, independent, honest, and defensible way. Thus the core of the Law of Decisions is the admonition to develop and employ your critical thinking skills.

For engineers, critical thinking is essential. Whether designing or improving a product, installing or operating a system, solving an operational problem, or, for that matter, managing a department or a corporation, engineers must use their heads—effectively. Besides their knowledge of basic subject matter, what engineers offer to employers and to society is disciplined brainpower. That brainpower is most effective if it is used critically and, especially, independently. That is, if engineers make up their own minds.

One of the barriers to effective application of the Law of Decisions is the existence of "authority." It is difficult to make up one's own mind when there is someone who is demonstrably well versed on the matter at hand and is able and willing to act as an authority in any discussions. When that person says (or implies), "I'm the expert here and this is the way we should do it," it is not easy to insist that you need a better understanding of the issue before you can sign off on the decision. But you must. You may need to say (more diplomatically, of course), "Look, if you're so smart, you should be able to explain to me the principles about which you are an expert." If they do, then you can make up your own mind.

There is a qualifier here, however. Engineering problems are so complex that it is not usually possible for everyone involved in a project to have a complete understanding of all of its aspects. You have to work as a team and rely on the expertise and the integrity of your fellow engineers. Then it becomes a matter of applying critical thinking to the decision of how far to go in accepting the judgments that are offered. Again, it is not easy to assess the reliability of the resident authorities, but it's part of an engineer's job.

The Law of Decisions is broadly applicable and I wish it were being applied more generally in the current political climate. It seems at times that the political advantage goes to those who can shout the loudest, not those who can provide a logical, critical analysis of the situation. In governing, as in engineering, the best approach is, "Let's think this through." And then make up your own mind.

While many whose academic roots are in the liberal arts lament the sad state of critical thinking in our society, perhaps it is engineers who are emerging as the critical thinkers. Indeed, if one examines the criteria by which ABET (Formerly the Accreditation Board for Engineering and Technology) evaluates engineering programs, one finds that programs must demonstrate that their graduates have:

- an ability to analyze and interpret data;
- an ability to identify, formulate, and solve problems;
- the broad education necessary to understand the impact of engineering solutions in a global, economic, environmental, and societal context;
- an understanding of ethical responsibility;
- an ability to communicate effectively; and

- a knowledge of contemporary issues.

Sounds to me like a pretty good definition of critical thinking. Perhaps ABET would do well to specifically mention critical thinking in the criteria.

In the end, the important thing is to make up your own mind and to do so objectively, reaching a conclusion through evidence and logic and not because you would like or not like certain conclusions.

32

DISCOVERABILITY

Don't record anything you don't want the whole world to see.

S OME YEARS AGO, I was on the board of directors of an organization that was experiencing a bit of internal strife. Since there was some possibility that this kerfuffle could result in legal proceedings, our corporate counsel gave the board a briefing on how we should comport ourselves. Among the things I learned in that session was a new word: "discoverable." In legal parlance, something is discoverable if it can be subpoenaed or otherwise brought to light of day and then used as evidence in a court proceeding. In particular, counsel warned us that anything we write—letters, emails, journals, or just plain personal notes—is discoverable. Well! The soft sounds of pens going into pockets filled the room.

There are various historical examples where damning letters or emails or tape recordings have brought the mighty low, but the potential liabilities of a personal written record had not really occurred to me before. After learning the legal meaning of discovery and reflecting a bit on the examples of recorded evidence that have resulted in considerable embarrassment—or worse—I am led to posit Lyle's Law of Discoverability: *Don't record anything you don't want the whole world to see.*

I know, I know. It was not so long ago that I presented Lyle's Law of Records: *Write it down.* At first blush, these laws seem to be in direct conflict. Not so. The difference is in the nature of the material recorded. The test is, would it embarrass you to have it widely circulated? Certainly you don't want proprietary laboratory notes to be seen by everyone, but it would not embarrass you if they were. But not every written—or photographed—thing will pass that test.

Which brings me to the phenomenon—I can't think what else to call it—that first brought this law to mind. Over the years, we have been treated to examples of love letters and compromising photographs surfacing at inconvenient times and in inconvenient circumstances, but their distribution has been somewhat limited in time and space. Today, technology has removed most such temporal and spatial limits. Social networking sites on the Internet make it possible for people whose technical capabilities exceed their social sophistication to post messages and photographs "just for their friends." But of course it doesn't work out that way. The whole world is their stage and they may find they have "friends" numbering in the millions.

Some people may beg to differ (my grandchildren, perhaps?), but I don't consider myself a technological troglodyte. I was using email back in the early days when you would call someone on the telephone to tell them you were sending them a message and then call them later to see if it had been received. And while I am not a fervid user of social networking sites, I am listed on one and have been able to use it to keep track of friends and colleagues. I have not, however, posted any nude pictures of myself (much to the relief of any potential viewers) nor sent any compromising messages or even recorded my innermost thoughts. My reticence is not due to my technical limitations, although I do have more than a few. I simply believe that whatever is out there is *out there.* Everywhere.

But the point of the Law of Discoverability is that everything is, or at least can be, everywhere once it is committed to writing, or especially once it is recorded in ones and zeros. It is discoverable either by intention or by happenstance. Columbus didn't set out to discover America.

Everyone occasionally becomes very irritated with someone else. When that happens, I have always advised the irritatee to write a long letter, memo, or email denouncing the irritator and telling them where they are wrong. Make it just as strong and nasty as you like—but don't send it. Let it lie on your desk or in your computer files for a couple of days and then discard it. Or at least revise it to soften and sweeten your words—since you may have to eat them—and then still delay a day or two before sending. That is no longer my advice. If you write the letter by hand and are careful to shred or burn it when you are finished, there is a reasonably good chance that it will not be discoverable. If it is committed to any kind of memory media, you should assume that it will never go away.

I am concerned that the Law of Discoverability could be interpreted as cynical advice to hide your peccadilloes. That is not the intent. The underlying message is that privacy is not what it used to be and that you don't have to be involved in legal proceedings to have your writings and photographs become public.

And there is more. The most important principle embodied in the Law of Discoverability is this: if you would be publicly embarrassed by these writings or photographs or drawings, perhaps you should be *privately* embarrassed by them. Perhaps they should make you uncomfortable even if you could guarantee that no one else would ever see them. If you don't want the whole world to see a picture or read an account of you doing something stupid, why take the picture or write the account? And even more to the point, why do that stupid something in the first place? Perhaps this incessant display of our innermost thoughts and our outermost bodies (some more outermost than others) will turn out to be a blessing. The fear of getting caught—while effective—is not a very noble motivation for not doing wrong. But if it makes us ask why getting caught matters, it will have served a noble purpose.

LOCALITY

Know your territory.

To People who have never traveled in the Midwestern United States, there would seem to be no distinguishable difference between Iowa and South Dakota. They are both located in a rather blurry region lying somewhere west of Pittsburgh and east of Lake Tahoe and they produce a lot of corn or something. And indeed, when one crosses the Big Sioux River from Sioux City, Iowa, into South Dakota, it appears that there is no difference. The terrain may be a little flatter and the trees may be a little sparser, but all in all, it looks just the same. But be not deceived. South Dakota is a big state. As you travel north and west, the flatness and the sparseness seem to increase gradually, and then, in the middle of the state, you reach the Missouri River. There you find a discontinuity.

As I understand the geology, the Missouri River marks the farthest extent of the glaciers, some 10,000 years ago. The effect of these glaciers is profound. On the northeast side of the Missouri, the land was scraped and rearranged in such a way that the tributary rivers run from north to south and they do so through relatively flat farm country. South and west of the great river, the land was not graded by the glaciers and the ancient tributaries run from west to east through rugged ranches and badlands. When my family moved west, lo those many years ago, we found that not only did the land-

form change, so did the culture. I felt a bit out of place in my base-
ball cap when all about me was a sea of cowboy hats. My tales of
squirrel hunting didn't seem to interest the guy (or gal) who had just
hung up a 200-pound mule deer. And who knew what "hung up"
meant? Before we really felt comfortable, we had to learn a lot about
this new culture. Thus, Lyle's Law of Locality: *Know your territory.*

The traveling salesmen in *The Music Man* sang it well, albeit with
questionable grammar: "You gotta know the territory." So what territory do engineers gotta (may I say "have to"?) know? Several. First,
no engineering work of any significance is ever done out of context.
Consider the laptop computer upon which I am currently compos-
ing—the brand of which I will considerately not divulge. I am sitting
on a stool at my shop workbench looking down at the keyboard and
all is well. The keys are shiny and reflective, but I can read them. (I
touch-type, but I cheat.) The screen shows me a reflection of myself, but I can ignore it. All is well. But I don't usually work at my
workbench. I usually work at my desk, where I am seated lower with
respect to the keyboard. In that case, the keyboard is angled such that
the screen—still reflecting my image—is reflected in the keys and
they are unreadable. The designers didn't give enough thought to
the salesmen's definition of "territory." It's where you sell things. It's
where your products are used, not where they are developed. In a lab,
where people sit on stools and touch typists don't cheat, no problem.
But out in the territory…there it would be much better if these keys
had a matte finish and the screen was nonreflective.

Another territory that everyone has to know is the place they
work. Every company, every university, sometimes every division of
a company, has a different culture. In my engineering career, I have
worked at five different universities and (coincidentally) five different companies. And they were all different. Their dress codes—or
lack thereof—were different. Their approaches to time of arriving
and time of leaving were different. Security varied from company to
company (and was generally stricter than at any university). It always

took a while to learn the culture—to know the territory—and then to adapt to it, but doing so was an important step in becoming productive in each position. So when going to a new job, get to know the territory. See what people wear—especially the people you want to emulate. Better to dress like your boss than the company clown. Learn where the substantive discussions are held. Are they in meetings? In the lunchroom? In the lab? Figure out how decisions are made. How much autonomy and initiative people are allowed to have. Go slow. Watch and learn.

In this era of increasing globalization, we all find ourselves exposed to cultures that are widely different from our own. If you find yourself assigned, even for a short time, to work in another country, it is especially important to get to know the territory. You can do a lot before you go, but you will still have a lot to learn once you get there. Probably your best ally in your early days abroad will be a heightened sense of humility. That and a few words of the local language will go a long way toward keeping you comfortable while you learn the local customs. A few years ago, my wife and I were in a little restaurant in Perpignan, France. I used my few words of atrocious French to apologize for my inability to speak their language and found that the waitress spoke reasonably good English. Across the room was a trio of Americans who not only didn't try to speak a little French, they made fun of the language. "Hey, I think I'll order the 'poison.'" Ho, ho, ho. Funny, but that waitress forgot every word of English whenever she crossed the room. Learn your territory, folks.

Now, learning the territory doesn't necessarily mean that you adapt completely to the culture, be it that of a company, a country, or a club. If the inhabitants of the territory behave in a way that offends your ethical standards, by no means should you conform. There may or may not be opportunities to *change* the culture, but if not, unto thine own self be true. In any event, first you gotta—oh, please—*you have to* know your territory.

Your job is greater than your assignment.

IN KEEPING WITH OUR HUMBLE ESTATE—and even more humble budget—my wife and I usually fly coach. Occasionally, however, an airline has a good sale or we cash in some long-hoarded frequent flyer points and enjoy the comforts of the business-class cabin, especially on a long overseas flight. Such was the case recently as we flew from Oslo to Philadelphia.

Forty thousand feet. Smooth ride. Comfortable seats. Good wine. And lousy service. I don't mean poor service. I mean *atrocious* service. The flight attendant repeatedly brought us the wrong orders after carefully writing them down. She would disappear for long periods of time. Our dessert and coffee dishes stayed on our trays for ten minutes…fifteen minutes…twenty minutes. Other flight attendants walked by en route to the galley, passing my used dishes without a sideward glance. "What is going on here?" I asked myself. And myself provided no answer. Finally, I stood up, took my own dishes to the galley, and found our flight attendant huddled with one of her colleagues, viewing photographs on his cell phone. I said—calmly, I think—"I'd like to see your supervisor." She replied —also calmly—"I *am* the supervisor." Well.

Then ensued a conversation, the details of which I will not burden you with except to say that it was calm and that I tried—I really tried—to be constructive and not too judgmental. Until... until I pointed out how many flight attendants had walked by without picking up my dishes and she replied, "But that's not their job."

Oh, my. I stayed calm—I really did. (Doth he protest too much?) But I did expound at some length on just what one's job should be before returning to my seat. Disgruntled as I was, however, I am grateful for this experience, because from that conversation was born Lyle's Law of Jobs: *Your job is greater than your assignment.*

Wherever you are employed, you have some defined responsibilities. Maybe you design jet engines. Maybe you write software. Maybe you teach control systems. Those assignments can be pretty specific or very general, but in any event you might feel justified in just designing jet engines or writing software or teaching control systems. Indeed, you can feel successful if you obtain excellent results in performing your assigned function. However, while I don't know much about management theory, I don't think an enterprise can succeed if everyone does nothing more than fulfill their assignments. One reason for this belief is that I don't think anyone can think of all the things that need to be done to ensure that success. One of the several reasons for the failure of communism is the practice of central planning, where the great planners define everything that needs to be done. A fatal flaw in central planning—among many—is that the planners can't think of all the tasks that must be done and then assign people to do them. Some will invariably be missed. Assignments alone will not assure success. Jobs will, if the jobs are greater than the assignments.

Like everyone, most of what I know I have learned from others. One of my best teachers was Harvey Fraser, who was president of the South Dakota School of Mines when I was beginning my career in academia. Harvey is a West Point graduate who fought in the Battle of the Bulge, later taught at West Point, and retired from the

Army as a brigadier general. He was a deliberate and decisive leader who wrought many improvements during his tenure at the Mines. He was fond of saying that a campus should be "clean, green, and serene." The "serene" part was particularly significant because that was during the turbulent sixties, but he was serious about clean and green, too. As he walked across campus at any time of day, if he saw a piece of paper or a discarded pop can (yes, "pop," we were west of the Mississippi), he would go out of his way to pick it up. Was that his assignment? Hardly. Was it his job? He thought so. And his example—not only, I hasten to add, in picking up trash—taught me to think so as well.

Would that piece of paper have been picked up if President Fraser had not done it? Of course. That was someone's assignment and they would eventually get to it. But in the meantime, "clean, green, and serene" would suffer.

It is also important for students to understand what their job is. One of the easiest and most effective techniques for improving the quality of education is, in my humble opinion, the use of clear educational objectives. Tell students what they should be able to do at the end of the course and they are much more likely to be able to do it. I'm sure of it. And what's more, people who know a lot more about education theory than I do are also sure of it. So, what's the problem? Well, one argument against using objectives is that if you specify what students are supposed to learn, they will learn only what is specified. I would like to reject that argument completely, but I have to admit that it has a certain amount of validity. But not if students understand that their job is greater than their assignment. It may be too narrow to say their assignment is to get a degree, but I don't hesitate at all to say their job is to get an education—in all its broadest meanings.

Perhaps the significant difference between an assignment and a job is one of motivation. Fulfilling an assignment is internally motivated: *I* want to keep my job. *I* want to get a raise. *I* want a promotion. Doing a job (as defined in Lyle's Law) is externally motivated. The campus looks clean and green. The company I work for is more successful. My community—or the community I'm visiting—is a better place. Someone else—a friend, someone I know, or someone I don't know—is happier or safer.

You will be judged by someone else on how well you do your assignment. You will judge yourself on how well you do your job.

35

LYLE'S LAW OF

REASONS

If you do things for the right reasons, you will probably do the right things.

CARD PLAYING HAS NEVER BEEN one of my favorite activities, but over the years I have enjoyed a few games of poker, hearts, and some others whose names I don't recall. The game of hearts is kind of fun because it is reasonably simple, involves both luck and skill, and requires a certain amount of strategy. For those unfamiliar with the game, the entire deck is dealt to four players who then play their cards and take "tricks." The object of the game is to avoid taking hearts or, especially, the queen of spades. Each heart counts one point against you, the queen of spades—which has several uncomplimentary nicknames—counts *thirteen*.

There are other features of the game that could make taking the queen of spades a good thing but usually, it hurts. One friend of mine, when someone stuck him with the queen, would always say, "Now why did you do that?" Why, indeed? Of course, everyone knew the answer. The sticker stuck the stickee to help the sticker win and the stickee lose. But my friend used the occasion to feign a whine and we all enjoyed a good laugh. That question, however, can provide a lot of insight, and should probably be asked more often, not to analyze what we did but to guide what we should do. To this end, I will posit Lyle's Law of Reasons: *If you do things for the right reasons, you will probably do the right things.*

So how do we decide which reasons are right? A few misanthropes might argue that it is not right to feed the hungry or to shelter the homeless, but most people would say those are pretty good motives. At the other end of the spectrum, it is hard to believe that the desire to get revenge or to punish someone could be considered to be noble reasons for doing something. Between these two extremes—feeding the hungry and exacting revenge—things can get pretty fuzzy. What constitutes a right reason or a noble motive? Philosophers have pondered and argued this question for centuries and I won't be so bold as to try to add anything to the discussion. That's not the point of this law, anyway. We don't need to agree on what is right or on how right something is, because the only reason for doing so is that we could then apply the law to someone else. That's not what this law is for. It is for you to apply to yourself.

So how do you apply Lyle's Law of Reasons to yourself? It is simple, but it is not easy. We are always told to ask ourselves, "Is this the right thing to do?" This law says that it is even more helpful to ask, "Why am I doing this?" Am I doing this for the right reasons? And you have to be brutally honest. Why are you *really* doing it? And I would suggest that if the word "I" or the word "me" occurs in the statement of your reasons, you need to take another long look. Not that you don't have to take care of your own interests. Of course you do. But there is a difference between self-interest and selfishness and only you can sort that out.

In our work and in our lives outside of work, we are faced with many choices. Some are easy—for instance, should I obey the law? Well, of course you should. But it is remarkable how often people can convince themselves that this is a stupid law and probably doesn't apply to them or to this circumstance anyway. And you know, there are times when a law should be disobeyed. Civil disobedience was at the core of what Gandhi and Martin Luther King and, indeed,

George Washington did. But those three—and many others—would have passed the test of the Law of Reasons. They had good reasons for doing what they did. The right reasons. And sure enough, what they did was the right thing to do.

Let me hasten to add that just because you are doing the right things, it doesn't mean you are necessarily doing things right. There are many paths that are paved with good intentions and we know where those paths lead. How often I have undertaken to do the right thing—and for the right reasons—and ended up botching the job. But correcting the problem of my ham-handedness is best left to laws authored by such luminaries as Peter Drucker and W. Edwards Deming rather than Lyle. I am comforted, though, by having tried to do the right thing for the right reasons. I would rather be criticized for not doing something well than for, well, not doing something.

If I may digress—and of course I may because I am doing the writing—I would like to suggest the application of the Law of Reasons to our criminal justice system. This system metes out various penalties, ranging from fines to public service to incarceration to death. I have often wondered what the system is designed to accomplish; *why* do we assign these penalties? If it is to remove the perpetrators from society so they cannot offend again or to teach the offender that such offenses will not be tolerated, those seem to me to be good reasons. But some penalties, it seems to me, are intended to exact revenge or just to punish because it makes us feel better. These reasons, I would suggest, are not worthy of an advanced society, which I hope we are. And if they are not the right reasons, the penalties we exact may not be the right things to do. It's worth considering.

As I write these words, I fear that I come off as preaching. That is not my intent. I am as frail as many and more frail than most, and certainly not qualified to preach. I have, however, thought at length about these things and want to share these thoughts with you. And I do believe I am doing so for the right reasons.

LYLE'S LAW OF

COMFORT

Beware the cozy comfort zone.

OVER THE YEARS, my wife has learned to be wary whenever I start a sentence with, "Honey, what would you think of …" And with good reason. For example, one of the times she heard those words was when I came home and said, "Honey, what would you think of spending a year in China?" To put this in context, we had been married something over ten years, six of which had been spent living in student housing, planning for the day when we could have a "normal" lifestyle with a house, a garden, pets, the whole works. Finally we had those things: I had a job as an associate professor and we had a home in the country, three children, two cats (one a little psychotic, but hey…), and even a garden. Now comes the suggestion that we suspend all this, pack up, and move to China (actually Taiwan) for a year. This would mean a change of plans. It meant that we would have to rent out the house, figure out how to get halfway around the world, find a school for the kids, and on and on. It meant, in short, leaving our comfort zone. We weren't sure we wanted to do that. But we did it. And we're glad we did.

Leaving your comfort zone is not an easy thing to do, but whenever we have done so, we have generally been well rewarded. Sure, there have been some bad moments. There have been some times when we were pretty uncomfortable. But in the end, we're glad we took the risk. For all of us, the comfort zone can get pretty cozy and leaving it requires some effort and some courage. But the cozy comfort zone can be a dangerous place, precisely because we don't want to leave it. Hence, Lyle's Law of Comfort: *Beware the cozy comfort zone.*

I am a member of an investment club and I enjoy going there and pretending I know something about picking stocks. Fortunately, I have picked enough winners that my fellow members put up with my pretense. Among the many things I have learned from my more skillful colleagues is an appreciation for the risk/reward relationship. If you want to enjoy the possibility of high rewards (return on investment) you must suffer the possibility of large losses. In other words, you have to get out of your comfort zone. A risk must be taken.

But there are different kinds of risks. One is the kind typified by the old joke about the famous last words of redneck drivers: "Here. Hold my beer. I'm gonna try something." That kind of risk is not very well thought through, but impaired judgment and a lack of information make it more acceptable and more palatable. That driver may have left his comfort zone but he didn't have a very good reason for doing so. In fact, the word "reason" may not even apply here.

Another kind of risk is the risk of getting caught doing something you shouldn't, like breaking the law. The folks at Enron took that kind of risk when they cooked the books, and some of their auditors did it when they looked the other way. Less dramatic—but no different in principle—is the driver who decides to drive 85 in a 55 mph zone. Or the person who cheats on an exam or on a spouse and risks getting caught. Again, they may be outside their comfort zone, but that's not the kind of risk we're talking about here.

So, if we agree to exclude the foolish risk and the risk of getting caught, let's look some more at the notion of getting outside that cozy comfort zone.

In the engineering process, we often need to get outside our comfort zones. Most real-life engineering designs require compromise, and compromise requires an element of discomfort. Any engineer would be most comfortable making a product that is 100 percent safe and reliable, but we know that can rarely be done. So designers need to leave their comfort zone—or perhaps more precisely, stretch that zone—to accept a design that is as reliable as possible while still possessing the other features it needs to meet the design requirements, like cost. Whether leaving the zone or stretching it, the coziness has to be overcome.

Perhaps the coziest comfort zones of all are the ones we create in our careers. It is easy to get into a position where we are good at what we are doing, we are amply rewarded and secure, and we feel we are making a contribution. What's not to like about that? But that's just the point—a zone can be comfortable simply because it doesn't contain anything that we dislike. That doesn't seem to be a very high standard for living one's life. But it sure can be cozy.

I write this law with some trepidation, because I know that more than a few of us are, as they say, of a certain age. The Law of Comfort should not prompt any of us to say, "I should have... I could have... I might have..." We did what we did. That's that. What the law should do is prompt our younger readers to ask themselves from time to time if they are getting too comfortable. It should be considered whenever an opportunity comes along that is a little scary, a little uncertain, a little uncomfortable. To take advantage of such an opportunity, you have to get outside your comfort zone. You have to leave the comfort zone to take a new job, to start a business, or, for that matter, to get married. But the chances are—and I really believe this—you'll be glad you did.

So beware the cozy comfort zone. It is a dangerous place because it is so hard to leave.

37

LINKING

Link authority to responsibility.

THOSE READERS WHO ARE OF MY GENERATION will recall that the late sixties and early seventies were a time of considerable ferment on college campuses. There were some weighty issues such as the Vietnam War, civil rights, the draft, and hardhats versus hippies. There were also some issues that were less cosmic in nature but were almost as intractable, such as dormitory visitation policies.

Before 1970, most campus policies governing visits to men's dorms by women and to women's dorms by men were pretty simple, boiling down to just one word: PROHIBITED. But a brave new world was emerging as students became restless and more liberated and the old principle of in loco parentis began to fade. Striving to accommodate this new reality, many campuses undertook a review of visitation policies. While our campus was by nature pretty conservative, our president had a good view of the cultural horizon and—somewhat reluctantly—appointed a committee to develop a new policy. Since I am telling this story, you have probably already concluded that I was a member of that august group.

Someone called us all together and we discussed and reviewed and rehashed and rediscussed and, after a few meetings, had gotten absolutely nowhere. Then one day I received a phone call from the president's secretary who told me the boss would like to see me. In his office, he asked me about our progress and I reported that there was very little. He then let it be known by word and gesture that it was my responsibility to get this job done, whereupon I replied that I was not the chair of the committee, whereupon he replied, "Well, you are now." So, newly burdened with responsibility but also armed with authority, I called a meeting of the committee, explained the new order, and we hammered out a visitation policy. I also learned a valuable lesson, which I state now as Lyle's Law of Linking: *Link authority to responsibility*.

You have undoubtedly heard or read that a particular manager is "good at delegating responsibility." That may be an admirable trait, but unless the manager simultaneously delegates the authority needed to get the job done, it is not really delegation. It is more properly called "passing the buck." If someone is to be held responsible for producing a particular result, it is essential they be given the authority over the processes that will determine whether the result is attained. This is another of those laws that one feels should be prefaced by the phrase, "Needless to say,..." but then you realize that it needs to be said.

Managers can start observing this law by agreeing that they will not claim to be "delegating responsibility." They will delegate *authority*. This is an action that requires courage and good judgment and organizational skill. The manager is saying to the other person, "I am taking my hands off these knobs and levers. You take over." The manager must then make it clear to that other person that since they now have control, they also have the responsibility of achieving the desired result. Authority has been delegated. Responsibility has been—what?—assigned? I don't think "delegated" is the right word. But whatever you call it, authority and responsibility have been linked.

An understanding of the Law of Linking is important to managees, as well. Whenever you are assigned a responsibility, you would do well to ask if the law is being observed. Ask yourself if it is clear that you have the authority needed to fulfill that responsibility. If you have doubts about it, see your boss and ask if you have analyzed the situation correctly, and, if you have, what an appropriate remedy would be. Of course, you have to do this tactfully so the remedy isn't just that you go to work somewhere else. But an astute manager will appreciate the fact that you are able to determine what you need to do the job and will make sure you have the necessary authority.

The Law of Linking can even be useful in child rearing. It has been said that the most important job of a parent is to create an adult who is no longer dependent upon the parent. That means giving children increasing responsibility as they mature. But you can't expect them to take on responsibility unless they have the related authority. You can't give your teenagers the responsibility of taking the trash to the dump if you don't give them the keys to the pickup.

While good leaders are adept at delegating authority and assigning responsibility, they recognize that delegation and assignment are not synonymous with divestiture. When a ship enters inland waters, she will generally take aboard a local pilot who then assumes authority over the navigation of the vessel and also the responsibility of seeing it safely through the various hazards to shipping. The captain, however, retains the authority to reassume control because he knows he cannot divest himself of the ultimate responsibility for the safety of the ship.

I saw this happen once when I was serving aboard USS *Norton Sound*. We came into Port Hueneme when the Santa Ana winds were blowing about fifty knots and the pilot could not control the ship. Unexpectedly, he suddenly appeared on the fantail, obviously unhappy. The skipper had assumed control and ordered him off the bridge. There were a few tense minutes but eventually we were safely tied up. I'm not sure what happened to the pilot as a result of this little drama but I am fairly certain that if we had crashed into another ship, our skipper would have lost his command and probably the rest of his naval career. No matter what happened, he was ultimately responsible.

It takes great courage to delegate authority. I would suggest that it takes even more to un-delegate it. But sometimes it needs to be done. Even here, though, Lyle's Law of Linking provides some useful guidance. If you take back the authority, take back the responsibility as well.

I should probably tell the rest of the story about the visitation rules. The code we developed was a compromise among the various factions, resulting in a copious compendium of rules that specified visiting hours, how wide doors had to be open, how many feet had to be on the floor, etc. But time continued its inexorable march and one year, maybe two years, later the whole thing was discarded and we pretty much deregulated dormitory visitation. In the process, Lyle's Law was obeyed. Authority and responsibility were tightly linked. Regarding this issue, the school now had neither.

38

THE LIMITS OF LOGIC

Think, but also feel.

MY FIRST ENGINEERING JOB was a summer internship at Collins Radio Company in Cedar Rapids, Iowa. It was a great experience and I have long said that I learned more engineering in those three months than in any other twelve-month period of my career. I worked for and with engineers and technicians who had been with the company for years and were very solid, both in their engineering fundamentals and in their dedication to their company and to their craft.

My assignment for the summer was to design and deliver a production test unit for a major component of a flight control system. In that short period of time, I had to learn the operation of the system, understand the parameters that had to be tested, work with other engineers to develop the test procedures and tolerances, and finally, design the test circuits. But it didn't end there; then I had to design the panel itself, choose the components, and get the whole thing fabricated and tested before I walked out the door. Can you see why I learned so much engineering?

What did I learn? I could write a thousand words to answer that, but I want to concentrate on a revelation that came very near the end of my tenure when I proudly showed the finished product to my supervisor. It was a rack-mounted panel full of meters, lights, switches, and knobs and it worked perfectly when I demonstrated it. When I was all finished with the demonstration, my supervisor complimented me and then asked, "Why did you choose those knobs?" Well, I guess I hadn't given very much thought to the knobs but they looked sturdy, they were easy to grasp, and they weren't very expensive. What's not to like? The boss then said, "Those are the ugliest knobs I've ever seen."

Wait a minute, I said (to myself). Sturdy. Easy to grasp. Cheap. What does "ugly" have to do with it? Well, as I thought about it, I had to admit that the knobs were not the most attractive. But still, simple logic would dictate I should use those knobs. And then it occurred to me—well, not then, but over the years—that engineering is not just pure logic. Good engineering certainly employs logic, but it also must, at times, involve feelings or emotions, intuition, a sense of beauty, ethics—a whole host of nonlogical parameters. Thence, Lyle's Law of the Limits to Logic: *Think, but also feel.*

Countless words have been written—and probably even more have been spoken—about the tension between logic and feelings in the leading of our lives. It seems that all of us listen to both; some of us more to one, some more to the other. Logic rules in one case, feelings in another. I trust that very few people consider themselves 100 percent logical. After all, we still beget children even though we know—if we think about it—that they will seriously restrict our activities for about two decades and also cost us a bunch of money. But, while we may lose money on the logic side, we are richly rewarded on the emotional.

It is less apparent, I think, that such logical ambivalence should extend to the practice of engineering as asserted by the Law of the Limits of Logic. But consider the nonlogical phenomenon we call intuition. I suspect we have all had the experience of solving a problem or working out a design or somehow reaching a conclusion and then stepping back and saying, "That just doesn't look right." No reason. Or at least no reason that we can put our finger on. But further reflection will often show that there really *is* something wrong or at least that something could be better.

What is this thing called intuition and how does it work? I don't know. Certainly it involves—or at least is informed by—experience, but it seems to go beyond that. Whatever it is, however it works, we are well advised to heed it. Many an error has been uncovered and many an improvement has been engendered by intuition.

Another nonlogical phenomenon is emotion, broadly defined as a mental state arrived at through feelings rather than reasoning or cognition. I ran across an interesting quote from Isaac Bashevis Singer, a twentieth-century Polish-American author who won the Nobel Prize in literature in 1978. Singer wrote, "The very essence of literature is the war between emotion and intellect." That may be true in literature, but I suspect that in engineering, the war is over— and intellect won. The vanquished, however, should not be purged or exiled. We need to integrate emotion into our postwar culture and listen to it from time to time.

My old boss didn't go through any reasoning process to arrive at the conclusion that those knobs were ugly. Looking at the knobs elicited in him a feeling or a reaction that was unpleasant and he knew he would feel better if he were looking at a different knob. Logically, the choice of knobs really didn't make any difference. Esthetically, emotionally, it did. It does. In our professional activities and in living our lives, it is our emotions that lead us to consider such things as beauty, ethics, justice, and love. To ignore those considerations will diminish our profession and our lives.

Another quote I like is from Rabindranath Tagore, an Indian poet and another Nobel Prize winner, "A mind all logic is like a knife all blade. It makes the hand bleed that uses it." Ouch.

While the Law of the Limits of Logic advises paying attention to nonlogical phenomena, it does not condone the use or even accept the existence of what might be called the "antilogical." By antilogical, I mean the know-nothing rejection of science and scientific knowledge that seems to be rampant in our society today. We can't eschew experimental science simply because we don't feel good about the results produced by that science.

All the while I have been writing this article, I have been trying to figure out how to get through it without reference to our viscera. Perhaps my readers will be more creative, but I haven't been successful. After all, when you are being affected by your intuition or your emotions, where do you feel it? I'm afraid I have to end by suggesting that you exercise your logic but that you should also listen to your gut. Sorry.

39

LYLE'S LAW OF

CONTESTS

Sometimes you win.
Sometimes you lose.

ONE OF THE ACTIVITIES IN NAVY BOOT CAMP WAS—and probably still is—instruction in various forms of self-defense, including formal gloves-on boxing. We may have followed Marquis of Queensbury rules but I think we mostly just flailed around while getting some of the basics of jabbing, blocking, bobbing and weaving, and, of course, ducking. As the one-week course came to an end, our instructor decided that we would have one final match between two "champions," one from each of the two companies that had undergone the training together.

For reasons still not completely clear to me, I had been appointed Recruit Chief Petty Officer of our company of some eighty young men. The RCPO of the other company was about my size, so the decision was made that the two of us would defend the honors of our respective organizations. The contest was not a pretty sight—at least not from inside the ring. As it turned out, we were both considerably better at offense than defense and as we approached the fifth and final round, the ratio of blows landed to those blocked became larger and larger. The fight ended when my opponent landed a powerful right—or maybe it was a left, I don't know—and good red Iowa-farm-boy blood gushed from both my nostrils and the fight was called. The instructor didn't declare a winner, but everyone knew. The guy sitting there bleeding had lost. Hence, Lyle's Law of Contests: *Sometimes you win. Sometimes you lose.*

Would I have liked to win that contest? Absolutely. Was I disappointed because I lost? Absolutely. Was I sorry I competed? Well, my nose had some regrets, but after a little time for reflection, I was satisfied. I had stood toe to toe with my opponent and had given as good as I got. The honor of Company 181 was intact. It was a fair fight. I would live to fight—metaphorically speaking—another day.

As long as there are contests, there will always be winners and losers. Certainly most athletic games are designed to exclude the possibility of ending in a tie. If three companies bid on a job, only one will get the contract. When twenty people apply for employment at a company that has three openings, there will be more losers than winners. The Law of Contests warns of this situation but it doesn't say what to do about it. That is because the advice appears to be so conflicting, I couldn't encapsulate it. Let me explain why.

The first bit of advice is to expect to win (see Lyle's Law of Expectations: *Model success. Expect the best.*), or, in other words, don't even *think* that you might lose. When you are writing a proposal, think about how you will execute the project after you have been awarded the grant. As you go into an interview, be making plans about how you will move into your new office. As you step into the boxing ring…well, here the advice is simpler: don't step into the boxing ring. But I trust that you get the picture. You *have* to believe that you will win and know that you will not lose. On the other hand…

On the other hand, you'd better have a plan B. A military commander moving troops into battle will always consider how to get those troops out if things go sour and they have to retreat. Prudence pays. So, if there are two agencies that might fund a proposal, submit it to both of them. (You can sort out the details after they both have awarded you the grant.) When you are looking for a job, don't just apply to one potential employer. In other words, plan what you will do if you lose.

So the advice is this: don't even think about losing, but think about losing. Contradictory? Not really. The first refers to your attitude: *believe* that you will win. The second refers to your actions: *prepare* for the possibility that you might lose. They are not mutually exclusive. I can believe that I have found a stock that will only increase in value but prudence dictates that I not invest all of my savings in it.

Finally, I hope the Law of Contests will remind you that losing is not the end of the world. It may hurt. It may inflict considerable damage. But if you have contested honorably, there is no shame in losing.

For some years there has been a cartoon on refrigerator doors and in office cubicles showing a very satisfied dragon picking his teeth with a jousting lance. Surrounding him are various pieces of armor, a broken sword, and a few horseshoes. The caption reads, "Sometimes the dragon wins." Sorry, noble knight, you have just become a casualty of the Law of Contests. Indeed, sometimes you lose.

But wait!—as the hawkers on TV say—there's more. The casual observer of this cartoon will probably overlook the battered shield behind the dragon. Emblazoned on the shield of the vanquished and now being-digested knight is the Latin inscription, *Veritas et Honor*, "truth and honor." The message seems clear: the knight is defeated and so, too, are the virtues he championed. Or maybe not. The shield is still there, a little dented and bent, perhaps, and unnoticed by the dragon as he enjoys his postprandial repose. But still there. I think the message of this cartoon is, "Sure, I lost this one, but as long as truth and honor and the other virtues are not compromised, I am not beaten." As Hemingway said in *The Old Man and the Sea*, "Man can be destroyed but not defeated."

Coming back to the boxing story, the final chapter was played out the next day when our two companies fell out to march to the mess hall. Even though I had bled profusely the day before, I showed up unmarked by combat. My opponent, on the other hand, had a black eye, and his nose was swollen and the color of an overripe plum, lingering mementos of our pugilistic encounter. Sometimes you win. Sometimes you lose.

40

LOVE

Love somebody. Love some thing. Love some place.

D URING **WORLD WAR II,** there was a concerted effort to induce people to drive less and thereby save gasoline for the war effort. Part of this campaign was the production of the "Victory Bike." a very basic but serviceable bicycle. My sister acquired one of these bikes with the plan that she would ride it to the one-room country school where she, at the tender age of eighteen, was the sole teacher. While that plan was largely unrealized due to the uncertain condition of the country roads, the bike served the family well. When the war ended, that sister moved out to marry her fighter pilot, my other sister moved out to seek her fortune in the big city, and my one brother who had not left to fight the war—as three others had —discovered internal-combustion-based transportation. That left me with the by-now dilapidated old bike. Better than nothing, but barely so.

In the months after V-J Day, American industry shifted with amazing speed to the production of civilian goods. Among the best of those goods—in the eyes of a twelve-year-old boy—were bicycles, and one day there appeared in the window of the local hardware store the most gorgeous bright-red Schwinn bicycle one could imagine. It was love at first sight. I discussed this affair with my father and he offered a bit of advice. "Never fall in love. It makes you do foolish things." I've thought about this advice a lot over the years and, well, sorry, Dad, but I think you were wrong. At least in part. While love may make you do foolish things, I believe that for our own fulfillment, we *do* need to fall in love. Indeed, I shall be so bold as to state, in three parts, Lyle's Law of Love: *Love somebody. Love some thing. Love some place.* Let's look at these individually.

Love somebody. It is hard to imagine anyone going through life without ever truly loving another person, but I suppose it can happen. Oh, what a bleak life that must be. Even Plato, a pretty practical guy, said, "And he whom love touches not walks in darkness."

Some of us are fortunate in having people who are easy to love. I love my wife. I love my children. I cannot imagine doing otherwise. But there must be people who don't have the good fortune of knowing the easy-to-love. And there are those who are unwilling to give up the self-interest that makes it difficult to love someone else. For these people, obeying this part of this law will be difficult but not, in my opinion, impossible. I venture, here, into waters I am not qualified to sail, but I believe that everyone is capable of giving love and will undoubtedly be the better for having done so. Remember the song from the musical *South Pacific*, "You've Got to be Taught to Hate"? Well, maybe you can be taught to love, too.

Love some thing. There are all kinds of "things." As a matter of fact, I guess there is nothing that is not a thing; we generally differentiate between things and people, although technically, a person is a thing. But I digress. In this part of the law I am not referring to the physical things—such as bicycles—that, while perhaps enriching our lives and expanding our experiences, can be generally categorized as property. I am thinking instead about what might be broadly classified as institutions: nations, universities, clubs, companies, churches, professions, to name a few. Sir Walter Scott wrote, "Breathes there the man, with soul so dead / Who never to himself hath said, / This is my own, my native land!" Saying that is an expression of pride, an expression of love for one's country, the kind of pride and love that anyone might have for their alma mater or company or any other institution.

How about our own profession of engineering? Do we engineers recognize that engineering is a profession with a long, proud history? That engineering is of paramount importance in determining the future of civilization? That we are bound by shared knowledge and shared ethics to other engineers around the world? If we do, and if we don't have "a soul so dead," we will have pride in—and love for—the engineering profession. And we will be better for it, and so will our profession. I fear that in our zeal for teaching our engineering students all the technical knowledge we feel they need to succeed, we fail to tell them about the glories of our profession, its rich history, and its relation to civilization. If we paid more attention to the broader picture of who and why we are, I think it would inspire a greater respect for and, yes, love of, the engineering profession.

Love some place. We live in a mobile society. While some people stay close to where they were born, many—especially the more educated—pursue a dream or an opportunity that takes them far afield. My wife and I have lived on both coasts and in the middle and in another country. Our children did not grow up where they were born nor do they live where they grew up. This mobility can be exciting and will certainly broaden our perspectives. There is a danger, however, that so much moving about can prevent us from identifying with any particular place. I believe we are the better if we have a place or some places that make our hearts beat faster when we return. One of those places may be "the green, green grass of home," but it is more likely, in this age of mobility, to be a place we have lived or visited or, most fortunate of all, where we now live. What could be better than to love where you live?

Some readers may think I have gone astray in going from professional and behavioral advice to the esoteric topic of love. But the goal of the laws is to help the reader to be more successful and, indeed, happier. With that being the goal, how can I neglect something so central to our beings as love? My father asserted that love makes you do foolish things, even though you don't really want to. Even here, I think he got it wrong. Love makes you *want* to do things that others might consider foolish. There is a big difference between, "I will do this because I love you" and "Because I love you, I want to do this."

In the end, Lyle's Law of Love is not only a law, but also my wish for you, my faithful readers.

Index

About the Author

LYLE FEISEL grew up on a farm in central Iowa and received his elementary education in a one-room country schoolhouse. After graduating from a small-town high school, he attended Iowa State University for one year, just long enough to meet the two loves of his life—his now wife Dorothy and the profession of engineering. When the money ran out, he enlisted in the U.S. Navy where he served three years at sea, helping to test the Navy's newly developed surface-to-air missiles. By the end of his enlistment he and Dorothy had married, and they returned to Iowa State where over the next six years they gained three children and Lyle earned three engineering degrees, including a Ph.D.

Leaving Iowa State in 1964, Dr. Feisel joined the faculty of the South Dakota School of Mines and Technology where he worked for the next nineteen years, teaching engineering and doing research. Those years were punctuated by brief periods of industrial employment at IBM and Northrop, adding to previous industrial experience at Collins Radio and Honeywell. The Feisels lived for a year in Taiwan where Dr. Feisel was a National Visiting Professor at Cheng Kung University in Tainan. They also took a sabbatical leave to fill the Wachtmeister Chair in Science and Engineering at the Virginia Military Institute.

In 1983, Dr. Feisel became the founding dean of the Thomas J. Watson School of Engineering and Applied Science at the State University of New York at Binghamton. There he reorganized and integrated some existing technical programs and developed several undergraduate and graduate programs in engineering. In his eighteen years at Binghamton he helped the school establish a reputation for high academic standards, outstanding research productivity, and service to the local community. Toward the end of his tenure, he was granted a four-month leave of absence, and he and Dorothy took a freestyle trip through Europe and around the world.

Dr. Feisel retired in 2001 but has continued to work with various societies and to serve as an adviser, consultant, and evaluator for universities in the United States and several other countries in Europe, Asia, and Latin America. He holds two patents and has published extensively on technical subjects, the theory and practice of education, and social and professional issues. He has received numerous honors and awards for teaching, service, and publication.

Acknowledgments

Over the course of the decade it took to produce Lyle's Laws, many people have made various contributions and are deserving of the author's appreciation. Certainly I must thank James Froula, former Executive Director of Tau Beta Pi and Editor of *The Bent,* and a longtime friend. In 2001, Jim recruited me to write a column for the magazine and suggested the name, "Lyle's Laws." His encouragement and his regular deadlines assured a steady flow of laws.

My wife, Dorothy, has provided support and constructive criticism of my drafts and has suggested several laws. She has also shown great patience while I was immersed in the task of writing and during long silences as I contemplated the best way to develop a law. Thank you, my love.

Our children—Patricia, Margaret, and Kenneth—have been especially supportive over the years. I thank them for that support as well as their years of forbearance as they grew up under the system of Lyle's Laws, long before those laws were committed to writing. I am especially grateful to Ken who has undertaken the task of editing and publishing this volume, a learning process for both of us.

Finally, I wish to thank that generally nameless group of people—parents, friends, employers, employees, coworkers, colleagues, fellow volunteers, hunting and fishing buddies, and more friends—who taught me, by good example and bad, the principles that are embodied in Lyle's Laws. I have gained much from them and hope they have gained even a little from me.

CPSIA information can be obtained at www.ICGtesting.com
Printed in the USA
BVOW02s0453110615

404147BV00002B/11/P